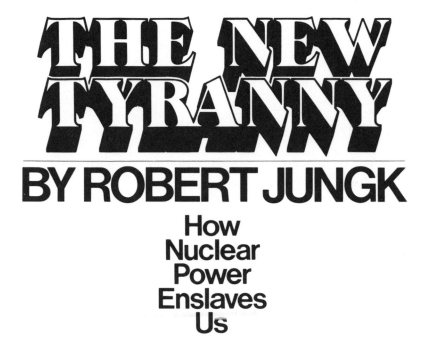

THE NEW TYRANNY

BY ROBERT JUNGK

How Nuclear Power Enslaves Us

TRANSLATED BY

CHRISTOPHER TRUMP

Fred Jordan Books/Grosset & Dunlap

A FILMWAYS COMPANY

Publishers/New York

Contents

The Hard Path

1

The splitting of the atom moved man into a new dimension of violence. What began as a weapon against one's enemies now threatens the hand that wields it. In substance there is no difference between "Atoms for War" and "Atoms for Peace." Nothing, neither stated intentions nor applied technology, can alter the inherent dangers of nuclear energy. Even its most ardent champions must admit that it is virtually impossible to eliminate *all* of the risks that this new form of energy poses. And it is these risks—however slight they may appear—that threaten catastrophe beside which all possible benefits pale into insignificance.

An atomic accident caused by technical failure, human error, or sheer malevolence could wreak havoc lasting decades, centuries, even thousands of years. Nuclear energy—and the fears that it might get out of control—now looms like a great dark shadow over mankind. It is there, whether we see it as an inescapable blight or as the ultimate destruction of all life on this planet.

This dark prospect must haunt all advocates of atomic

1

energy. Yet they continue to insist that we can find refuge in effective safety measures no one has yet been able to devise. If it were only a matter of technology we could quibble over engineering problems and the economies of cost. But the plain fact is that nuclear energy demands that we be made safe from man himself—his mistakes, weaknesses, rages, his cunning and lust for power. Protect nuclear facilities from these foibles and passions and you run the risk of a regimented society that would appear tolerable only in contrast to the dangers it seeks to avoid.

All of us, nations and individuals, must confront this stark reality. For too long we have been concerned with the biological and ecological effects of nuclear power while ignoring its political and societal fallout. This book forces the issue. It was written in fear and anger: fear because of the threat to freedom and mankind; anger against those willing to barter our highest values for profit and creature comforts. It will be faulted for shedding more heat than light; for violating the comfortable dogma of "business as usual." But to approach the power of plutonium with calm discourse is to make the awesome danger it represents appear trivial. There is simply no substitute for outrage when the fate of mankind hangs on the slender thread of reasonable projections based on false assumptions.

One of these long-held assumptions is that nuclear energy is all right so long as we avoid war. But today we are confronted with the prospect of terrorism, theft, or atomic sabotage that could bring disaster to a nation regardless of external enemies. Would terrorists exercise the restraint of statesmen? In light of recent events, the question begs no answer. After Hiroshima and Nagasaki there were widespread demands for stringent atomic arms controls. The proliferation of "Atoms for Peace" poses the new problem of avoiding atomic civil war.

The hazards of nuclear energy know no bounds. The dream of infallible domestic security is a myth. Even if we turn nuclear industrial states into the equivalent of concentration camps we can never be sure of avoiding nuclear blackmail and violence. It is just as real as the threat of international nuclear blackmail. And a police state would be especially vulnerable to internal rivalries, with the inherent danger of one side making the "ultimate threat" against the other. The ruthlessness of dictators has inevitably led to disaster; in a nuclear state that final reckoning may well mean oblivion.

2

Amory B. Lovins is a sensitive young intellectual who looks like a bookworm, yet spends part of each year living somewhere in the northeast United States as a camp counselor. In the fall of 1976 an article Lovins wrote for *Foreign Affairs* on the dangers of nuclear energy caused considerable stir among experts in the field. Since then he has traveled throughout the world, carrying his message to businessmen, top government officials, and scientists that they stop pursuing the "hard path" of energy escalation. Lovins tries to show them that projections for future electrical power needs are totally unrealistic; the quantified illusions of their own hopes and fears.

According to Lovins, false projections based on erroneous calculations and goals unleashed the "atomic euphoria" of the sixties. He sees it all ending in a massive hangover simply because the ambitious schemers of the nuclear industry and its lobby will be thwarted by technical, economic, and political considerations.

But Lovins is more than a mere naysayer. He urges a gradual return to a "soft path" of decentralized alternative energy sources less destructive to the environment. Small and medium-sized plants would serve to create more jobs and give citizens greater control of a vital resource. Such a technology would also narrow the energy gap between developed and developing nations. However, Lovins admits that what he proposes is "against the interests of a number of powerful institutions."

An examination of the industrial and political development of the last century reveals that the majority of societies has opted for the "hard path," dismissing the "soft path" as backward. The conclusion is depressing. The decision for nuclear energy was the logical result of a political thrust that ruthlessly pushed productive capacity ahead of all other human concerns.

The present struggle against nuclear power cuts across class lines, uniting people of all nations who reject this burgeoning technological monster and all of the stress, destruction of nature, and potential dangers it represents. The "hard path" has reached both its zenith and its breaking point. We now recognize that it has concentrated power in the hands of a few, widened the gap between the "haves" who can never get enough and the "have nots" who grow ever poorer because they know not how to help themselves. It is a line that leads to oblivion—to a chilling estrangement, isolation, and belligerency.

For now, those following the "hard path" are determined not to let the facts stand in the way of their policies. They often answer their critics with force and suppression, which serves only to heighten the violent tensions unleashed by the nuclear era. It threatens to pollute not only our earth, but our political and social wellsprings in the process.

Even more worrisome is the fact that the so-called socialist countries have followed the capitalist world in pursuit of the "hard path" in nuclear energy. In these countries, dissenting views which might bring about a shift in policy are, for the most part, suppressed. Indeed, it is likely that the once much-touted adaptation of the two systems to each other's ways will more likely result in western acceptance of the socialist world's repressive measures. Already nuclear energy supporters here have expressed their admiration for the "discipline over there."

We in the "free world" have seen a sharp increase in intolerance, direct or indirect censorship, the hounding of "dissidents," and government intrusion in our professional and private lives. Many express the hope that these are only "temporary measures." But a country opting for nuclear energy is choosing a "strong state" in perpetuity.

Put bluntly, nuclear energy provides the justification for the power elite of industrial nations to pursue "tough policies" and a "hard path." Those who do not submit to such an authoritarian form of government are simply dismissed as "subversive."

In this setting, government and the economy function like a huge machine that has variously been described as "grinding down" or "crushing" or, again, "consigning to the garbage heap of history" those who oppose it. Nuclear energy technology exported to the Third World serves only to strengthen already existing authoritarian forms of government, never allowing democracy a chance to develop and driving the rural population from the land. Nuclear energy makes economic sense only if it is produced in large centers and distributed from there. In the developing nations many industrial cities already too bloated for their own good are thus encouraged to grow still further. This conveniently assembles a large mass of "customers" in one place. Supplying the villages, on the other hand, would make necessary a big and costly distribution network. This explains why only a fraction of the energy from Indian reactors reaches the rural population which needs it the most.

On a broader scale, we face the prospect of nuclear energy imperialism, with dangerous and highly vulnerable "nuclear parks" generating huge amounts of electricity at selected sites in Africa, Asia, and Latin America. In time, more and more as yet independent nations will be drawn into the "nuclear web."

The ultimate horror of the nuclear scenario for Third World governments—and others as well—was sketched out by Albert Wohlstetter, an American political scientist. In his plea to stop the spread of nuclear weapons, made at the 1978 Windscale Hearings in Great Britain, Wohlstetter said: "Implosion weapons in the kiloton or greater range . . . might be used by governments as a desperate last resort against populations." At the time Wohlstetter made these remarks it was assumed that he was referring to such countries as Iran and South Africa that were confronted with "crisis management" in the face of powerful minority pressures. But just what was it then that New Hampshire's Governor Meldrin Thomson had in mind when, during the height of the Seabrook crisis, he said that he wanted to arm the National Guard with nuclear weapons? Such is the abyss to which the "hard path" could lead us.

3

An exaggeration? Perhaps, since this book could not be published if matters had already come to such a pass. But the hour is late. Even now totalitarian technocrats are trumpeting that "the future has already begun."

One of the peculiarities of nuclear development is that while it is difficult to stop at the outset, once the reactor has started its action is irreversible. The waste it generates will be with us for generations to come, permanently guarded to protect everything that lives from its lethal radiation . . . decades, centuries, thousands of years. As the number of sites to be guarded grows, so must the power that mandates these controls put its mark on our political structure. Once we have firmly set out along the "hard path" there is no turning back.

Today an aroused consciousness prompts more and more people into opposing policies that will permanently affect future generations. They realize that whatever the short-term benefits of nuclear energy—doubtful as these are—they could not possibly outweigh the monstrous burden it would place on man and the environment for all time. It must be assumed that the peaceful uses of atomic power can be shielded effectively from human error and malevolence *only* then when all people are locked into a "new tyranny."

Those who protest are condemned as "crusaders"—as if

they were blindly seeking salvation in some outmoded technology. The truth is that many opponents of atomic power have thought long and hard before pitting themselves against the disastrous course being pursued both by those in power and those elected by the majority. Numerous studies have shown how nuclear energy endangers human life. This book will show how it also threatens our basic freedoms.

There is still a chance to get off the "hard path." But it is a slim one.

ONE

Radiation Fodder

1

"If someone stays in the 'hot zone' longer than I've allowed, I simply cut off his oxygen supply. That doesn't leave him much choice. He doesn't dare take off his headgear to get some air, because everything in the room is contaminated with radiation. So he comes out. Fast."

A bright, intelligent Frenchman, Patrice Fleury, tells me this. Fleury is a conscientious young man when he is not too tired to think. As a radiation monitor in the *Section de Protection contre les Radiations* at La Hague nuclear fuel reprocessing plant Fleury's job is to make sure that employees are protected from excessive doses of radiation. He says he hates being *un sale flic*—"a lousy cop." His job of constantly watching and warning is not only thankless, but practically impossible. New leaks are a chronic menace in what he calls *cette boîte pourrie*—"this stinking hole." They spew out radioactive poisons that instantly contaminate hair, skin, eyes, lungs—a real danger to the workers were they not covered from head to foot in plastic shrouds.

On paper the reprocessing plant is designed to function

with a minimum of human intervention. The procedure includes:

- Unloading the fuel rods from special carriers
- Storing the rods in water to reduce their radioactivity
- Slipping off the protective coating with remote-controlled arms
- Cutting up the rods with huge, remote-controlled shears
- Dissolving the fragments in hot nitrous oxide
- Separating the uranium from plutonium and other elements
- Making plutonium oxide and concentrating the uranium
- Treating the waste material and separating it into liquid and solid storage units
- Burying the wastes according to the degree of radioactivity
- Discharging slightly radioactive liquid wastes into the sea

In practice, however, the process is an obstacle course fraught with innumerable pitfalls. Boilers exposed to powerful acids and intense heat have buckled; pipes have burst and valves sprung leaks. To this day no one knows for sure why such an unusually large number of leaks and material breakdowns have occured at La Hague—or at other nuclear plants throughout the world. If everything worked out as the planners had expected, men like Patrice Fleury would have an easy job of it.

But, as the Swedish physicist Hannes Alfvén pointed out, the nuclear power industry is more a nightmare than the dream its supporters proclaim. Neither machines nor men have proven as efficient as the blueprints drawn up by technocrats. Man is careless, forgetful, inaccurate. He is subject to fatigue and daydreaming. In short, he is not a machine. Yet nuclear technology demands superhuman precision and accuracy. This dichotomy is seldom so obvious to an observer as here at La Hague, on the northernmost tip of the fog-enshrouded Cotentin peninsula of Normandy.

This is the site where the French Atomic Energy Commission (CEA) has erected the world's largest reprocessing plant for nuclear fuel. Its chief function is to recover plutonium (Pu-239), the precious artificial element extracted from fuel rods that have expended only a fraction of their energy potential in atomic reactors. Plutonium is the stuff of bombs and fast breeder reactors—the next generation of nuclear technology.

Nowhere in the world have these reprocessing plants worked satisfactorily. Continual breakdowns have led to stoppages and permanent shutdowns, as at the West Valley reprocessing plant in New York State. Though experts admit that reprocessing technology is too experimental for large-scale use, huge plants have been set up. Those at La Hague and Windscale in Great Britain are designed to handle hundreds of tons of nuclear fuel. Round the clock huge trucks shielded with lead—the French call them *châteaux* (fortresses) —roll along the country roads of Normandy from Germany, Italy, Holland, Sweden, and Spain, bearing to La Hague the lethal cargo that every nation wants to get rid of.

Undeterred by frequent breakdowns, accidents, strikes, and the growing unrest of local residents (all of which French authorities do their best to conceal from the outside world), representatives of the *Compagnie Générale des Matières Nucléaires*

(COGEMA) scour the world in search of contracts to nuclear fuels. The latest fat deal was signed with Japan. Ever since West Valley was forced to shut down and Windscale was unable to accept further contracts, the French have held a monopoly on nuclear fuel reprocessing.

For years no news of the breakdowns at La Hague leaked out. Politicians, businessmen, local government officials, journalists—all carefully selected by the French atomic industry—were invited on inspection tours. They were shown impressive factory buildings dominated by an enormous chimney more than 300 feet tall. They visited shops and control rooms where work was progressing smoothly but were hurried past those areas that were shut down once again for repairs. The guests were treated to generous helpings of hearty Norman cuisine, washed down by *calvados* apple brandy at convivial dinners. There the plant's managing director assured them: "We have had no protests from the few people living on the Cape."

Yet even a cursory glance at local newspapers would have revealed how alarmed local residents had become. Bernard Laponche, a physicist working closely with the French atomic authority and a leading official of the *Confédération Démocratique Française du Travail,* a predominantly social-democratic and Christian trade union representing the majority of the organized workers at La Hague, denounced the claim that La Hague is working well as "a lie." No one listened, not even after Laponche exposed what he called the "cover-up of La Hague" at several press conferences late in 1977.

If the truth about La Hague were known, the captains of nuclear industry in several countries would have an infinitely more difficult job. No longer could they claim before their respective government agencies that the storage and reprocessing of atomic wastes posed no serious problems because of current contracts with France and La Hague. What would happen if this dumping place of the last resort were no more?

Thanks to Monsieur Laponche, I steered clear of the COGEMA public relations staff and made direct contact with the workers who daily risked their lives at the plant. They had no interest in showing visitors some phony façade of an atomic Potemkin village. They want the world to know what is really going on. For example, in the summer of 1977 the plant's cooling basins had built up dangerous levels of radioactivity because the rods slated for reprocessing had been stored in them too long. Breakdowns elsewhere in the plant had held up the entire process, making it impossible to achieve the planned daily output of four tons. As matters stood, the plant was unable to process fuel even from French reactors, let alone the material arriving every day from abroad. A new addition to the plant in 1976, capable of processing ten times more material than the original plant, had been out of commission for more than a year.

The workers at La Hague gave me a glimpse of the world's most frightening working conditions. They have sacrificed not only their health, but their rights to free speech and self-determination. They refer to themselves as "radiation fodder"—the cannon fodder of the new technology. All fear that after a few years of working there they will be so much "waste" dumped on the unemployment rolls or, worse yet, the hospital. They have no faith in any sort of compensation for claims they might make years later for the delayed consequences of radiation overdoses. To date the record seems to confirm their pessimism. Just as the Americans showed no willingness to pay anything in compensation for the delayed effects suffered by the victims of Hiroshima and Nagasaki, so La Hague's management has failed thus far to accept long-term responsibility for radiation-linked diseases among its former employees.

2

At the end of his shift Daniel Cauchon falls into an exhausted sleep on the plant bus that takes him from the work area to the parking lot. He awakens only after the bus has cleared the guarded area. For years Daniel has been working in *intervention mécanique,* repairing defects discovered by radiological safety teams. According to the planners, mechanical failures were supposed to be the exception. In practice, repairs—both minor and major—are the order of the day. The bugs had no sooner been cleared out of the main plant, UP-2 (*Usine Plutonium 2*), when old age seemed to set in. The contractors, apparently in a hurry to finish, had done a sloppy job.

"For one thing, parts didn't match. Nothing fit anything else," veteran workers at La Hague told me. "It was maddening. At the time we hoped that eventually things would improve. We are still waiting, but no one believes it any longer. It was even worse in 1976, when the new American-designed plant for fuel rods from light-water reactors, the *Atelier HAO (High Activity Oxides),* was put into operation. It had to be shut down after a few weeks—and hasn't worked since."

A breakdown in a reprocessing plant is infinitely more

complicated and time-consuming than in a conventional factory. Highly toxic sources of radiation must first be isolated. Entire sections of the work area must be closed off for hours or even days before a leak can be plugged, a bent pipe section replaced, or a break repaired. In a major break the units must be taken apart piece by piece and decontaminated before being put back together again. It is work for an atomic Sisyphus, made all the harder because the burden being pushed up this new summit of technological supremacy is extremely poisonous. The strain takes its toll not only in the physical demands of the job, but also in the mental stress caused by the fear of contamination. And this fear is made worse by the oppressive isolation and restriction of movement imposed by the cumbersome protective suits—the armor that is supposed to protect them—which men like Daniel Gauchon must always wear when doing their jobs.

La Hague workers call their protective suit a "Shaddok" —after a mischievous bird-like cartoon character on French television whose long beak resembles the pointed gas filter on their face masks. At first, these shrouds of nuclear *haute couture* were called "Hiroshimas" or "Nagasakis"; but the analogies proved too grim, even, for Gallic gallows humor.

It takes close to a half-hour for a "plutonium knight" to don one of these "Shaddoks" correctly. Under the watchful eye of the radiation safety staff he first puts on white underclothes, followed by a skintight white garment with a red stripe around the chest area. Over this goes a coverall of vinyl. In addition, the worker must wear three pairs of special socks and overshoes, three pairs of gloves, plus a breathing mask covering both mouth and nose. Over all this goes the "Shaddok" itself, another pair of special gloves, and a final attachment: an air tube that the worker drags along beside him like an umbilical cord.

Once ready for work, the worker is briefed on the duration of his stay before entering the airlock on the way to the "hot zone." Depending on the level of radiation, this may be hours or only minutes. Also considered is the individual's exposure record as listed on his individual radiation balance sheet. If in the preceding month the employee absorbed a large percentage of his annual allowable quota—which by French law is ten times greater than for workers in non-nuclear facilities—then his stay in the hot zone could be brief. Indispensable specialists are used sparingly, so that they can remain available to supervise difficult jobs and other emergencies throughout the year.

Some repairs take hours to complete, so that often three, five, or ten workers spell each other until the job has been finished, each worker only doing a part of the job. Many workers find this to be intolerable. They have to get used to the idea that they can never finish a job, that all they are allowed to do is only one segment of it. Because they never see the beginning or the end of their efforts, they feel all job satisfaction is denied them.

In 1969 a damaged container at the Saint-Laurent-des-Eaux power station took 14 hours to repair; some 105 men worked on it in shifts, each receiving substantial doses of radiation. In 1970 no less than 700 welders were required in the United States to repair a single broken pipe in a steam generator at the No. 1 Indian Point power plant that supplies New York City with electricity.

La Hague and other atomic installations have hit on a highly questionable "solution" to reduce the radiation exposure to which their highly skilled and highly insured manpower is exposed. They hire part-time workers (*intérimaires*) from nearby employment agencies, called "slave dealers" by the French. The latter keep the radiation records of the *intérimaires* in a manner that is far from diligent. For example, it is simply assumed that these workers have never before been exposed to radiation at other nuclear facilities. No one bothers to check.

These men are generally given the "dirtiest," most dangerous assignments. As the scab workers of the atomic industry they often receive in a few days as much radiation as a regular employee does during a whole year. They are the first ones sent into a contaminated area to do the preliminary work for skilled employees, such as closing leaks, setting up entry locks around the leaks, and putting contaminated clothing and radioactive waste into plastic bags for disposal. They are asked to hold their breath as much as possible to avoid stirring up radioactive dust.

Often these "temporaries" receive more than the allowable dosage of radiation because their agencies simply "forget" to send in the radiation film badges required by health authorities. Also, having little or no training, they are more likely to be injured (such as being burned by direct contact with radioactive materials) than are regular employees. All too often they land in the infirmary on their first day of work. During the holidays students, who learn quickly but have little manual dexterity, are often used to do this kind of work. But most of the time the *intérimaires* are drawn from the ranks of the unemployed, lured to the job by high pay. They are not told how dangerous the work at La Hague can be.

Such exploitive practices have prompted trade unions to

demand "radiation passports" and safe working conditions for the *intérimaires*. Management has not responded favorably, since to do so would probably mean shutting down the plant. Since 1967 the contamination rate has been soaring, making it impossible to continue without using workers who are either uninformed or desperate enough to endanger their health by ignoring safety regulations. The price they must pay isn't noticed for a while—we know the effects of contamination take years to emerge. Blind to the future, they hire themselves out for an hourly wage which will later cost them years of their lives.

Regular workers also grow more careless as they continue working at La Hague. Perhaps the constant threat of radiation danger has inured them to the hazard. Sometimes they avoid putting on protective clothing to do a minor job in a contaminated area. Why bother putting on the entire rig with its many individual parts which are just as cumbersome to take off again? It is simply too much trouble and takes too much time. The "Shaddok" is a monstrosity that makes you sweat; your hands tremble and your heart pounds in your throat. The oval mask on the helmet fogs over so that you can neither see nor hear your fellow workers. Reduced to sign language, you feel utterly isolated. You are nagged by fear that you will bump into something with your clumsy outfit or else tear it on a sharp edge.

A torn "Shaddok" means a headlong retreat to a "clean zone," undressing so hastily one is prone to making a wrong move that can lead to truly serious radiation damage. Such contamination would mean hours, days, or even weeks of painstaking examination in the medical section. Everything would have to be analyzed—blood, saliva, mucus, urine. The worker inquires anxiously: "How much did I get, doctor? Is it serious? I've been pretty tired and irritable lately . . . my wife says I'm not fit to live with. And our sexual relations haven't been the same, either."

In time men learn to outwit the safety controllers. They sneak a smoke where it is prohibited, nip a bottle of beer where drinking and eating are banned. They figure out how to avoid the hassle of being classed "over-dosed" by fiddling with the small, fountain-pen–shaped radiation counters all are required to wear whenever one of the frequent spot checks takes place.

These breaches of discipline with such a terrible potential cannot be attributed solely to carelessness or indifference. They are something of a revolt against the intolerable spying and controls of the plant. Eventually many of the workers come to feel that the whole routine is little more than harass-ment forced on them by paper-shuffling managers, aloof engineers, and dictatorial radiation-monitor men.

Patrice Fleury confessed to me that even the radiation safety controllers have long since tired of the constant, thank-less vigilance forced on them by their jobs. They are too few to ensure that all the rules are always obeyed. And if they enforced them by the book, the operation would have no al-ternative but to shut down. Often they will look the other way if, say, the night-shift workers improvise a gripping tool to fish for defective fuel rods in the storage basin. Otherwise, it might take too long to wait for the repair of a broken remote-controlled arm. Yet if this makeshift device loses its hold, the result will be a spray of contaminated liquid over everything. It would be recorded as still another "incident" that could close down that particular section for days.

"Perfect safety exists only on paper," Fleury told me. "You hear about it only at scientific conferences and in the industry's sales literature. Cutting out improvisation would mean even lower productivity. By using their own ideas, the men have the feeling that they are more than just cogs in a machine. The brochures put out by COGEMA have color photos of the most magnificent apparatus. What they fail to show is the most important tool of all—adhesive tape, which is used to fix just about everything."

Despite the efforts of public relations people to cover up breaches of safety regulations, the official record shows that from 1973 to 1975 the number of contaminations rose from an annual 280 to 572. Workers claim that the figure today is still higher, except that the statistics are now no longer publicized.

However, it was impossible for La Hague officials to cover up one such "incident." Three inspectors from Euratom, an agency that oversees nuclear development in Europe, happened to be on the scene two years ago when a leak in the plutonium shop forced up readings on safety gauges 30 times beyond permissible limits. Thereupon, in the summer of 1976, the unions demanded that the entire plant be completely decontaminated and modernized.

The plant administration responded with delaying tactics. Then came the rumors that the French government was planning to turn over La Hague and all other nuclear facilities to private industry. Labor unrest grew. Workers feared that under private ownership the pursuit of profits would lead to dangerous new production norms and further delay long-needed safety improvements. They also worried that the relatively high pay rates would be leveled off to those paid elsewhere by private industry and that social services would be cut. The result was a strike at La Hague in September 1976.

3

Banner headlines carried the story of the first sit-in by workers at a nuclear installation. The occupation of the plant lasted only twenty-four hours until management got a court order to remove the interlopers. But the strike itself went on for nearly three months, giving workers a chance to tell their story publicly about hazardous conditions at the plant. Even relatives and neighbors of the "radiation fodder" were shocked to hear for the first time what went on behind the walls of the mysterious plant. Demonstrators in "Shaddoks" flickered like silvery ghosts across French television screens. But somehow COGEMA was able to prevent these news reports from circulating in the foreign media.

The strikers organized concerts, solidarity picnics, and a hearing that lasted several days called *Les Assises du Nucléaire*. (In French this phrase has a double meaning—"the nuclear sit-in" or "the nuclear criminal court.") The strikers also published a satirical newspaper with a caricature of Adolf Hitler on its front page that looked suspiciously like their plant director. Hitler announced that if he were still alive he would join COGEMA and get elected its *führer*.

"This strike was much more than an ordinary labor dispute to us," Daniel Cauchon, one of the leaders, told me. "It was something of a big celebration. We had crawled out of the darkness into light. For the first time we met colleagues and saw that they were not robots, but comrades and friends.

Before that we only talked when necessary—we were always too exhausted, too tense to care. Then, too, there was that constant bickering between departments, making things tough all around. Now even some of the office staff seemed sympathetic toward us. We really had two objectives in common: to stop the dehumanizing working conditions and to prevent the takeover by private industry."

The apparent solidarity began to crack, though, when management offered new contracts to individual workers, giving them a deadline date by which they had to sign. The four trade unions involved began to squabble among themselves, with the communist-led *Confédération Générale du Travail* the first to break ranks. Although it was opposed in principle to La Hague's being handed over to private industry, it rationalized that workers' rights could be safeguarded only "on the basis of reality." All the same, none of this would have defeated the workers if they had not been brought to heel by the *vidanges* (drainage) incident, the implications of which loom larger today than the strike itself.

After several weeks of neglect because of the strike, the water tanks with their stored fuel rods had become so dangerously radioactive that they threatened to contaminate the entire area. It was feared that the gases generated would lead to an explosion, as had happened once before in 1973. At this critical juncture management asked that the men who normally flushed the tanks interrupt their strike to do this job once more.

Tension ran unusually high. Several plant trucks had been damaged by striking workers who were unhappy with the progress of talks with COGEMA in Paris. Already Daniel Cauchon and another striker, both advocates of nonviolence, had begun a hunger strike in the village church at Octeville. In this strained atmosphere the strikers refused to cross the picket lines. Pushed to the wall, management asked the government to declare a state of emergency. Put another way, the striking experts were "drafted to save the plant." The strike leadership had to comply, admitting to itself that the strike had never been a total one—some men had remained and ensured the safety of the installation throughout the walkout.

Society was suddenly faced with a new dimension in labor unrest, thanks to the *vidanges* incident. An atomic plant cannot simply be shut down like any ordinary factory. Physical processes are at work all the time and cannot be stopped without the most dire consequences. For example, highly toxic elements that could endanger both the plant and the surrounding countryside could be released if the cooling system were switched off or some of the equipment run at reduced capacity.

The French sociologist Christiane Barrier-Lynn told me about a similar incident during the strike at the nuclear power plant at Chinon on the Loire. One of the negotiating workers reported back to his colleagues after a particularly depressing meeting with management: "It was damned difficult. The boss flung his 'xenon effect' business at us and said we risked contaminating the whole area if we cut back any further on the output." An angry young worker protested: "Let's just walk out on this mess! If the area is contaminated maybe people will realize how much they need us." He had not been there long and had to be calmed down by older workers who explained that in a nuclear plant the right to strike had its limitations. Radioactive processes never stop; they have to be supervised for decades, centuries, indeed thousands of years.

4

The strike in La Hague was over in December 1976 with only a partial success. The workers could not prevent transfer of the plant into private hands. But they forced the creation of a commission of unions and management that made no less than 47 recommendations to improve working conditions. Based on this report the *Service de Sûreté Nucléaire* (Nuclear Security Board) insisted that the opening of a new plant, the first of its kind to handle fuels from foreign light-water reactors, be held in abeyance until the safety of workers and the environment could be guaranteed.

In the meantime the old UP-2 plant lurched on from one breakdown to the next. Management tried desperately to catch up on time lost during the strike. The result was still more mishaps, the worst occurring in Zone 817 where plutonium oxide, the most valuable end product of the chemical separation process, was treated and weighed. During four weeks early in 1977 there was only one incident-free day. The alarm was sounded no less than 41 times during this period. Work areas had to be evacuated five times in a single week because radiation contamination of the floor had reached such dangerous levels.

It was like a nightmare. Decontamination squads scrubbed day and night in what seemed an endless project. The number of workers treated in the infirmary for receiving excessive doses of radiation soared.

More trouble occurred at discharge installation No. 44.

Despite the fact that it was a violation of plant rules, workers were ordered to remain on the job after 5:00 P.M. No radiation supervisor was on hand to stop it, nor to prevent dangerously contaminated water from suddenly welling up from clogged drains. Rough concrete floors, the most difficult to clean, became contaminated. After the area was cleared it took 11 days to clean up. It was precisely incidents such as this one, previously hushed up by management, that now aroused the people of the Cotentin peninsula.

"They lied to us. They lied to us again and again! Now we no longer believe them," say the peasants and fishermen of Cap de la Hague when the conversation turns to the *Goul' Hague* (alluding to the Soviet Gulags) today. Years ago they thought they had made the shrewdest deal of their lives in selling the land on the windswept bluffs. Before the reprocessing plant was built the scrubby soil was virtually useless. Why not sell it to a company that was going to make television sets—or so they had been told. In fact there were a lot of buyers; some talked of making washing machines, refrigerators, and other appliances. A welcome diversification to the area's meager farming and fishing income—and, if one dared mention it, to the centuries-old smuggling trade with England.

Times were hard in the face of the big agricultural operations and modern fishing fleets elsewhere. Here, at least, was the promise of jobs for young people. The old timers, as always, were suspicious of all *horsains*—those whose families had not lived for generations on the Cotentin peninsula. In this land of the druids still dotted with their huge, upright stone monoliths or *menhirs,* the elders recalled the ancient prophecies. Once the druids had correctly foretold the ancient Viking invasions. Hadn't they also foretold a future invasion by land?

When the people on the peninsula first heard rumors that the building going up on the 540-acre site had something to do with atoms, they got nervous and went calling on the mayor and town council. They were given reassuring answers. The rumors were just that; nothing but rumors. Then came the builders, the suppliers of services, the engineers—and, finally, the first scientists from Marcoule, the plutonium production facility in southern France that had produced France's first nuclear weapons.

Although the truth could no longer be denied, its impact could be softened. The people were told that the plant posed absolutely no danger; it would be both quiet and clean. That tall chimney? Not to worry; it would never smoke and fume, but merely carry off waste gases. Those long pipes going into the sea? Only harmless wastes would flow through them, to be dispersed quickly by the Raz Blanchard, that tumultuous current that eternally swept round the peninsula.

Indeed, the Raz Blanchard was one of the main reasons the peninsula had been chosen as the site for a reprocessing plant. For centuries it had spelled doom for many a Cotentin fishing boat. Now it would serve to wash away the most noxious of pollutants from French shores. The site, already selected by French authorities in the late fifties, was also blessed by strong winds that supposedly would disperse the mildly radioactive gases. A non-porous subsoil seemed ideal for medium-term storage of nuclear waste. There was one final consideration: In case of disaster the peninsula, surrounded on three sides by the sea, could be effectively cut off from surrounding areas, something impossible with an in-land site. It was a contingency that was never mentioned publicly, of course, but its implications could not be ignored.

Today La Hague's management admits that con-siderable quantities of radioactive gases and liquids pour out of its pipes and chimney, which, with its delicate brickwork, looks like a model in a toy building set. (Despite this charm-,ing aspect, its function poses a real threat; tests conducted around a similar plutonium plant on the outskirts of Denver have shown a significant increase in the incidence of cancer.) The pipes leading into the sea have cracked at least 30 times, spewing radioactive liquid wastes onto the soil which, in some cases, has seeped all the way into the ground water. The sea as far out as 100 kilometers (60 miles) from La Hague now registers above-average levels of radioactivity.

Tests made in the Bay of Ecalgrain (into which La Hague wastes flow) show radioactive readings in marine life five times higher than at Cap Fréhel, 100 kilometers distant. Fishermen tell of finding wartlike spots on perch, sole, and halibut. In other fish the flesh has turned black. And as hap-pens so often when no one really knows what is going on and yet fears the worst, there are wild rumors of such things as sea monsters with multiple heads and tails being hauled up in nets. These alarming stories quickly make the rounds with devastating impact on the local economy. La Hague crabs, once considered a delicacy, are now practically impossible to sell. The famed *beurre de la Hague* (La Hague butter) had to be renamed *beurre de Val de Saire* because so many people re-fused to cook with it. Once, in the late sixties, the govern-ment had to buy up all the milk from the surrounding area because so much iodine-131, which affects the thyroid gland, had escaped into the atmosphere.

Such accidents are coming to light more and more often. A spot check on a truck belonging to the largest grocery store in Cherbourg revealed radioactive contamination. The day before, it had been rented out to transport some radioactive material needed for research. The containers had been improperly sealed. Debris smuggled out of the plant—it still had on it the three violent triangles on a yellow background that serve to warn of radioactivity—was found in a garage in the neighboring village of Valognes. Small animals and birds moving over and under the fences around the plant carry traces of contamination with them.

Children out picking wild berries suddenly fall ill. Is it radioactive poisoning? Tests show that the area immediately outside the plant has somehow been contaminated. Was it through underground springs? No one knows for sure. In time, the people living on this lovely Norman cape will come to fear their environment—every leaf, blade of grass, insect, breath of air could prove deadly.

At long last the politically conservative population of the cape began to stir. A Committee Against the Atomic Pollution of La Hague was formed. Assisted by a group of Paris-based scientists who are critical of the French government's nuclear policy, the committee has shown that in many places radioactivity has exceeded legally acceptable limits by 10 to 20 times. At first the government disputed such findings— only to have them confirmed in a second test done in the presence of a notary public. The committee has posted on walls and telephone poles everywhere stark yellow posters that read:

```
┌─────────────────────────────────────────────────────────────┐
│                                                             │
│   OFFICIAL STATISTICS FOR THREE YEARS (1972–75)             │
│                                                             │
│  Canton de Beaumont (the area immediately around the plant) │
│                          203 cancer victims per 1,000 deaths│
│  Cherbourg Arrondissement    185 cancer victims per 1,000 deaths│
│  Saint-Lô Arrondissement     163 cancer victims per 1,000 deaths│
│  Coutances Arrondissement    155 cancer victims per 1,000 deaths│
│                                                             │
│                         WHY?                                │
│                                                             │
└─────────────────────────────────────────────────────────────┘
```

The political well in the area is also being poisoned. Despite repeated assurances from the government that no additional nuclear sites would be built, present plans call for six more reactors to be installed on the Cap de Flamanville, widely known for its majestic cliffs.

Once again the local inhabitants were not consulted in an open manner. The first approaches were made through a series of confidential briefings with the mayor and the town council. When a vote was taken, only one voice was negative. A referendum held four months later resulted in 425 votes for the reactors and 248 against. It is said that the outcome was assured because men recently laid off by the closing of an adjacent ore pit were promised jobs in the construction of the facility and later at the plant itself.

The owner of the town's only café had become a particularly vocal advocate of the plant. Long-known in the area as a tight-fisted miser, he suddenly began setting up drinks for all "on the house" in the weeks just before the referendum. Who paid for them? The *pronucléaires,* supporters of the project, began using the bar as a meeting place. They harassed opponents of the project, calling them *"pollués"* ("the contaminated ones") whenever they met them, such as Didier Anger, his wife, and children.

Anger, a friendly, courageous elementary school teacher, is leader of the opposition. "People no longer talk to one another," he told me. "They swear a lot at each other and often come to blows. Farmers who resist selling their property to *Electricité de France* are beaten up on dark roads by young thugs. We suspect they get paid for it. The nuclear plant is supposed to create jobs."

In early 1977 machines were brought in to test the granite under the soil, an illegal move since the property had not yet been acquired under eminent domain. Two hundred peasants turned out to protest. When gendarmes led three of them away, the rest of the demonstrators set up roadblocks and occupied the site. They held out for nearly a month in their rock fortress, until driven out by 250 armed troops. The latter were stationed on the fields and cliffs night and day, angering even those who had supported construction of the nuclear plant. "It's worse than the Occupation," said one of them. "Instead of German troops, we now have German shepherds. Those dogs are trained to kill and their teeth are damned sharp. All you hear nights is their constant barking."

No longer are local people allowed to walk on their own fields, except with a special pass. One of them, Victor, a genial goatherd, refused and had many an argument with the guards. "Do *you* have an official order authorizing you to do this?" he asked. They shoved him away with their rifle butts. He stubbornly demanded official confirmation of these "measures to maintain order." Victor took his protest all the way to Cherbourg, where he was finally given an orange-colored pass that permitted him to cross his own fields. Angrily he tore up the pass and tossed it away, thought better of it, chased after the pieces and pasted them together. Now he is entitled to walk across the construction site, even though under the law it does not yet exist. "I'm pulling out," Victor said recently, "who wants to live like this?"

The guards withdrew temporarily to permit demolition crews to dynamite six million cubic meters of rock needed to gain more land from the sea. Some of the farmers joined together to prevent purchase of the 99 acres of land still needed by the power company in addition to the 54 acres ceded to it by the municipal council. But the prices offered are so high that some cannot resist. One farmer wept with shame when he told his family of his "surrender." Those that remain have banded together into the *Groupement Foncier Agricole* (Agricultural Lands Pool). They lay claim to 61 acres and say they will refuse to sell. At this writing the French government is making plans to confiscate their real estate.

The farmers have worked hard to keep the issue before the public. They have sponsored festivals and debates. Demonstrators from all over France keep arriving to show their support. Often they will stare across the barbed-wire fence at the blue-uniformed guards, who in turn stare back at them. At times tensions mount and angry words are exchanged. At one point the demonstrators seized a member of the local council who had been a particularly vocal supporter of nuclear energy and painted him green. The result: more police and an immediate ban on all public demonstrations. On one occasion police dogs were released against the demonstrators. And in nearby Siouville, a sleepy, family-style resort on the coast, an eighty-room high-rise with a fine view of the sea is being built. No, it's not going to be another hotel—but a police barracks.

5

Madame Lemonnier often wakes up with a start in the middle of the night. Although she lives several kilometers distant from the plant at La Hague, she strains her ears to catch the low, persistent hum of the exhaust fans. When the night wind carries it to her at Jobourg, she sinks back on the pillow with relief. Since that night in May 1977 when the sound suddenly stopped, she has lived in constant fear for her husband's life. Everyone had to flee the plant on the night the fans stopped. It could have been worse. If the airlocks had failed, the entire plant might have been contaminated. Without the cooling of the exhaust fans the deadly mixing baths could have overheated dangerously. There might have been an explosion to scatter radioactive material over a wide area. Fortunately the power was restored within six minutes. But the people around La Hague live in fear that, next time, they might not be so lucky.

Their greatest worry is the *Fossé Nord-Ouest,* a burial pit for the most dangerous, long-lived fission material just 600 yards distant from the main plant. Behind heavily guarded gates stands a massive, squat building large enough to admit transport trucks with their lethal loads. Once inside, the radioactive waste is dumped into steel containers 300 feet beneath the surface. Here it will be stored in the earth for thousands of years, like some dormant, brooding monster. Except that it is not dormant.

Close by, one can hear a hum. It is the sound of an air

conditioning plant, operating 24 hours a day, 365 days a year for centuries to come, cooling off the waste that can develop 1,300–1,500°F unless constantly cooled. Without the air conditioning the steel containers would melt, causing a major disaster.

This may well be what happened in 1957 in a not yet fully explained explosion in the Soviet Union. It is believed that a military nuclear waste dump in the southern Urals blew up, severely contaminating several hundred kilometers of the surrounding countryside. Lev Tumerman, formerly a leading scientist at the Moscow Institute of Molecular Biology who has since emigrated to Israel, reported that when he visited the area between Chelyabinsk and Sverdlovsk, he saw abandoned villages and wrecked buildings. They had been leveled by the government to prevent the inhabitants from returning to them. Woods, fields, and lakes were so heavily contaminated with cesium-137 and strontium-90 that the entire area will likely have to remain under quarantine for generations to come.

The world, of course, knew nothing about this. Even Soviet nuclear experts profess ignorance of the incident, despite the fact that the head of the Soviet Atomic Ministry was later accused of negligence for letting the facility get "out of control." Clearly, the atomic age is made to order for totalitarian states. Such governments are not plagued by citizens' inquiries or protests. Thus, an immense disaster like the one at Chelyabinsk can be totally blacked out of the news. Curiously, there was one loophole. Numerous articles in Soviet scientific journals at the time discussed radioactive contamination of plants, soil, animals, and the atmosphere. They served not only as evidence of the disaster, but helped western experts pinpoint where it had occurred. One of the articles by Zhores Medvedev, a dissident Russian scientist, was printed in *New Scientist,* a British weekly.

6

When I visited La Hague the Medvedev report was on everyone's lips. People were concerned that it could happen *here.* Would the entire Cotentin peninsula be turned into a huge atomic wasteland?

The people of Beaumont, Beauborg, and Auderville wondered why sirens had recently been installed on the roofs of their town halls. What should they do if those sirens went off? They asked about *Orsec Rad,* the evacuation plan rumored to have been drawn up in secret by the authorities. As always, the answer they received was the same: silence.

It was one of those misty days so typical of this part of Normandy when I visited the plant. I was standing at the atomic cemetery where most of the radioactive waste was buried, a scar of yellow clay gouged out of the green countryside. On it stand hundreds of gray concrete barrels into which the steel containers will eventually be placed. They are surrounded by a low fence that could easily be climbed. Only dosimeters attached to the fence posts to detect ambient radiation give any hint that there is something unusual about these storage barrels.

They go deep into the ground, deeper still than the Celtic necropolis believed to be buried on the site and which at one time was supposed to be excavated. Now it will never be dug up because the ground is too contaminated with radioactivity. Millions of liters of radioactive materials are buried here, dangerous forever, yet contained in tanks that will corrode in time. It is extremely difficult and dangerous to empty them at regular intervals, but it must be done if the ground water is not to become polluted.

Significantly, Professor Emilyanov, a Soviet scientist with whom I discussed this problem, was far more talkative than he had been earlier about the still-classified Chelyabinsk incident. He said that Russian scientists were still baffled by the problem of nuclear waste disposal. To illustrate his point, he told me that Soviet geologists had recently discovered what they thought to be a rich new source of uranium in the Ural Mountains. But when the location was excavated they found no uranium—only ordinary rock. Yet the Geiger counters continued to register strong radioactivity. Further investigation revealed a secret nuclear waste disposal site 40 kilometers (25 miles) distant. The tanks had leaked, spreading radioactivity to distant points through natural subterranean water channels.

The new factory planned for La Hague will experiment with a technique to vitrify—or seal in glass—radioactive wastes. Yet even within these glass blocks the radioactive elements will remain active for thousands of years to come. No one can predict whether flaws will not develop 10, 15, 50, 100, or 1,000 years after they have been buried. With what certainty can one say that startling geological changes will not take place in the land over this extended period of time?

"We were promised that the wastes buried in steel containers will be left there only temporarily," a young engineer told me as he took me on a secret visit to the atomic cemetery at La Hague. "But I don't believe it. It will be much too dangerous and difficult to move. Right now the foreigners who have dumped their garbage on us are paying us handsome fees. But will they continue to do so for decades, for centuries to come? Will they take it back? I simply can't believe that."

His vision of the future was far more pessimistic than the official projections announced by management. He spoke quietly, hesitantly, often with long pauses; not at all as fluently as I am trying to set it down here. He loves this corner of France and continues working at the detested *Goul' Hague* only because this is his home.

"But how long will people go on living here?" he asked. "The new high yield oxide plant, which was in operation for only a few weeks in 1976, will—when it is finally reopened—process on its own 800 tons of fuel annually." He predicted that by 1990 "we could be processing 3,000 tons a year. And by 1995, when the fast breeders are on line, we're going to have stuff that is by far more toxic than anything we have had to date at the UP-2 plant. Can you imagine how many *fuites* (radioactive leaks) all that will involve? Or how much radioactivity will be poured into the environment and into the Channel?

"Soon the six reactors near Flamanville will be operational and Pechiney-Kuhlmann [a leading chemical corporation] are already planning a uranium enrichment plant a little further south. In this way the whole damned fuel cycle will be in close proximity. This is going to be the most toxic place on God's earth—if it isn't that already! Soon the entire landscape will be studded with high-tension wires. Have you seen the plans for the trunk lines from Flamanville to Cherbourg and on to Paris? Two or three cables side by side with metal pylons more than 40 meters

(130 feet) tall. Huge slashes across the countryside. We don't need all that power here. It's going to Paris. They need a lot of electricity in *la ville lumière*!

"I have an inside view of the problem," he continued, "and I don't think there is any way they can ever resolve the risks inherent in reprocessing and disposing of the nuclear industry's raw materials. Inevitably, one by one, the plants will deteriorate to the point where they will have to be abandoned—radioactive mausoleums to our 'progressive' age. There needn't even be a catastrophic explosion. The area will gradually turn into an atomic wilderness. It will be so contaminated that, within a few decades, the entire cape will have to be hermetically sealed off and abandoned as a total loss.

"When that happens, the last human beings will finally have to leave *Goul' Hague*. Like some enormous source of infection it will have to be guarded for decades and centuries to come. The guards will be people who won't understand how we ever allowed this to happen. These people, whom we will never know, will curse us in our graves."

TWO

The
Gamblers

1

The Four Horsemen of the Apocalypse—pestilence, war, famine and death. In our age they have been joined by another: industrial catastrophe. This fifth rider may yet exceed all the others in its potential for destruction.

Throughout history natural disasters—the so-called acts of God—have eventually been forgotten. The wounds healed. But this is not the case with man-made catastrophes. A major accident at a chemical factory, biological facility, or atomic plant can compound the immediate trauma with awesome consequences for future generations. This explains the profound fears of people likely to be affected by such occurrences—anxieties that simply do not exist among those living, say, in the path of hurricanes or in an earthquake fault zone.

I first came across this anxiety when I visited Hiroshima in 1957. The *hibakushas* (survivors of the first atomic bomb) told me that as much as their own lives were altered by the exposure to radiation they received "that day," they were even more fearful of the harm done to generations yet to come. They were even more worried about what was to become of their grandchildren and great-grandchildren than about their own damaged existence.

The frightening shadow of long-range poisoning has cast a pall over our hopes for future generations. It has spawned a deep-seated guilt that has altered our perception of human existence. This is a direct result of our new knowledge—and use!—of atomic fission. For years no one shared my view; then I met Robert Jay Lifton, a psychiatrist who teaches at Yale University. He visited Hiroshima after I had been there, but talked to the same people, holding in-depth interviews with 75 *hibakushas*. Lifton's interviews revealed conclusively that these people were scarred mentally as well as physically. The bomb had profoundly shaken their faith in the future, their expectation of living on through their descendants.

Nearly a decade later, I came across a scientific study on the psychological impact of *non-military* nuclear energy. Its author, an American named Philip D. Pahner working under a contract from the International Atomic Authority of Vienna and the International Institute of Applied Systems Analysis of Laxenburg (Austria), wrote in his paper, "The Psychological Aspects of the Nuclear Controversy":

> The nuclear power plant is an actual and potential threat of destruction on a scale that is simply unimaginable. It assails an individual's perception of his life, its meaning and its future. The strain of living with this knowledge may well undermine the creative processes of the individual and society.

Pahner's findings, in striking contrast to those of nuclear

proponents who held that such anxiety neuroses did not exist, made me eager to seek him out. He had left Vienna. His mother in California either could not or would not tell where he was. Even Pahner's closest friends did not know what had become of him. It would appear he had dropped out of the world, or at least the nuclear scientific community to which he had belonged.

One of Pahner's colleagues gave me what may be a plausible explanation: "Everything that Philip revealed in his paper was completely at odds with the ideas of those who gave him the job. They assumed he would simply dismiss the mental trauma engendered by atomic technology as insignificant or as the product of traditional superstitions regarding technological progress. But Philip's study raised doubts. It showed that nuclear energy could not be compared to the railroads or the beginnings of the industrial revolution in the nineteenth century. Deeper fears were felt that simply could not be dismissed. Contrary to the prevailing practice, his contract was not renewed. Philip took this very much to heart. I think he was greatly disillusioned because his carefully considered conclusions were found intolerable. They did not agree with the prevailing mindset."

2

The International Institute of Applied Systems Analysis is one of the most interesting "think tanks" of the postwar period, inspired by McGeorge Bundy, one of the "best and brightest" of the Kennedy advisers during the sixties. Located in Laxenburg, Austria, in a Hapsburg castle that once housed Napoleon's unfortunate son, the duke of Reichstadt, the Institute seeks to explore on an international basis the promise and the dangers of industrial civilization. Cooperation between scientists of the so-called socialist nations and the capitalist west is much closer than had once been anticipated. They all are bound together by a shared ideology: an overpowering faith in technological progress. They defend this dogma against all skeptics. Although the Institute's first director, an American named H. Raiffa, was of an open mind, the driving force soon revealed itself as the deputy director, Professor Wolf Häfele. The son of a country parson, Häfele is the most outspoken promoter of nuclear energy. He must have found Pahner's gloomy vision of a nuclear future quite heretical.

A graduate of twelve years' research at the Karlsruhe nuclear research center in West Germany, Häfele ranks as a tireless propagandist, both among his colleagues and in the public eye, for that most dangerous of all reactors, the fast breeder. In so doing he has made a number of faulty predictions. For example, back in 1969 Häfele assured skeptical German authorities that the SNR 300 breeder reactor at Kalkar would be operational in short order. It is still not ready; and, in fact, is not likely to be so before 1983.

In the early sixties when sponsors of the SNR 300 sought financing from the government, Häfele estimated costs at "only" 165 million marks. Toward the end of the decade his estimate had risen to 500 million marks. Since then cost estimates have soared to 3 and even 4.5 billion marks. This monumental miscalculation prompted Kurt Rudzinski, a leading German science writer, to accuse promoters of the fast breeder reactor project of playing Russian roulette.

Attacks such as this have not deterred Professor Häfele

from proposing ever bolder visions of the future. He is a prime example of the reckless high roller in technology, using his considerable influence to invent grandiose new projects. As a group these promoters, though often scientists themselves, bear little resemblance to the painstakingly responsible, patient scholarly researchers of the past who helped transform man's concept of the universe in a short few hundred years. Instead they are scientific impresarios skilled at manipulating governments and economic leaders to support their hazardous extravaganzas. In their lust for power they are more akin to industrial magnates than to true scientists.

Professor Häfele often shares his visions about his "global strategies" at meetings of his "inner circle." Participants at these gatherings are expected to manage a visible display of enthusiasm. Debate and contradictions are clearly undesirable, whether the boss is plumping for an absurd scheme of building fast breeder reactors at the base of Austrian alpine glaciers or indulging in his dreams of a planetary government in which the "new technology and a new social order enter into a symbiotic relationship."

Central to this glorious future is the fast breeder—the most dangerous of all reactors because of the vast quantities of plutonium it produces and the ever-present danger of an explosive chain reaction. Professor Häfele has done more than any other European physicist to bring a nightmarish "plutonium world" into being. His vision scares even such ardent supporters of nuclear energy as Sir Brian Flowers and Edward Teller, "father of the hydrogen bomb." The recent decision of the French, Italian, Dutch, and Belgian governments to pool their resources to build fast breeder reactors is due in large measure to the enthusiastic lobbying of this one man. It has brought them on a collision course with the White House and its own profound reservations about breeder reactor development.

Häfele's rhetoric deals in the grandiose. He compares construction of nuclear reactors to the building of Europe's great cathedrals; a technology that bears witness to the genius of our age. One who has heard his tirades confided to me that he "positively intoxicated our scientists. Today's new Führer would talk just like that, mixing a heady brew of technological progress with a mystical sense of mission." Häfele, well aware of his charisma, has been known to boast of his skill at captivating audiences, including leading churchmen. Said he of his own speech to Protestant leaders at the thousand-year anniversary celebration of Loccum Cloister: "In their enthusiasm they fell flat on their bellies before me!"

Häfele dismisses objections that breeder technology is still in its infancy and that its sociological implications are awesome—something most supporters of nuclear energy concede—with the simple dictum that one must live dangerously. That is how he sweeps aside the time-honored practice of painstaking trial runs for new technical installations before they are put into regular service. He tells laymen that major and complex technical systems like the fast breeder cannot be tried out in their totality. Therefore, he says, we must be content with tests of their component parts. "It is precisely this interplay between theory and experiment, or trial and error, that is no longer possible for new technologies which are designed to master unique challenges," Häfele admits. "In the case of reactor safety this is obvious. It is not acceptable to learn reactor safety by trial and error."

Unfortunately Häfele is not alone in departing from the unwritten ground rules of technological innovations. Under these rules, no new apparatus was allowed to leave research and development without first having been thoroughly tested in its totality for possible failures, and under conditions which kept it in careful isolation from the environment. Today, new reactors are put into operation in densely populated areas without experimental knowledge about the unpredictable interplay between thousands of components which make up these complex gigantic systems. Computer simulations are now substituted for actual trial runs. Mathematical data representing component parts of nuclear installations are fed into tapes and possible or hypothetical mishaps are then assumed with the hope of gaining some idea of the possible course of dreaded misfunctions. In this way the inhabited world is becoming a vast experimental lab for the most dangerous technology ever devised.

Keith Miller, a mathematician at the University of California in Berkeley, took part in one such test. He expressed extreme skepticism as to its reliability. In an interview over CBS-TV on May 12, 1976, Miller said that the computer programs used by the United States Nuclear Regulatory Commission (NRC) "are nowhere near as complex as the real problems." The results were, he continued, "about as reliable as tomorrow's weather forecast. I wouldn't risk anyone's life on a weather forecast."

But Wolf Häfele is not easily deterred. He feels there is a price to be paid for everything. He accepts the residual risk as the price for the "extraordinary benefits" we might expect from nuclear energy.

Häfele confidently basks in his philosophy of the

monstrous risk. He even managed to turn a symposium of the World Council of Churches, a body more than a little skeptical about the benefits of nuclear power and fearful of the dangers of the fast breeder reactor, into a forum for his bold advocacy of risking the lives and health of others. In his book *Facing Up to Nuclear Power,* published by the Council, Häfele claims that mankind is now at the point where decisions affecting the future of civilization could and should be based on theories that need not be proved conclusively in advance. He deems it to be sufficient that these theories possess a high degree of probability. Moreover, he stresses this pioneering role assumed by nuclear industry in mankind's advance into the uncharted territory of the unknown.

Modern research thrives on the experimental confirmation of imaginary leaps of the inventive spirit. To this approach we owe many of the discoveries of our modern age. But never before have the risks been so great as those posed by the immature projects of nuclear energy. Lucky breakthroughs of the recent past that were based on imaginative and risk-fraught visions have emboldened the gambling fringe of today's researchers and engineers to the point of recklessness. It is a way of thinking that carries in it the seeds of catastrophe.

These methods were first spawned in the weapons laboratories of World War II, to be tested on major military objectives. "Thinking the unthinkable" (as Herman Kahn put it) became the fashion during the cold war of the fifties, with a morally and ethically unthinkable atomic war claiming hundreds of millions of victims plugged into computer models as an "acceptable" option. Researchers staged elaborate electronic war games that "took into account" the destruction of entire nations and continents. This gave rise to an entire generation of scientifically trained gamblers oblivious to the inhuman implications of their models. At first confined only to military scenarios, their methods have entered the civilian sector and now find credence and application in government planning at all levels, including the decision-making of the industrial complex.

All of these experts have one thing in common. They ignore public input; the people are no more consulted than are soldiers in time of war. Those who dare speak out are dismissed as uninformed or, even worse, as disloyal. Surely, the experts say, the vision of an atomic age must not be thwarted by cowardly critics evading their civic responsibilities.

Politicians fare no better when they speak up for the interests of their constituents. If they don't yield to the "greater expertise" of the scientific community, the latter churn out reports attacking their scientific credibility. The juggernaut of the nuclear establishment rolls over political initiatives with the assurance of scientific inviolability, at best making only minor concessions to those who object.

Professor Häfele has made it a point to apply the "war games" methodology to the public sector. No doubt he was guided in this direction by his teacher, Professor Carl Friedrich von Weizsäcker. I still remember the first time I met the latter in the mid-fifties at Göttingen. He immediately described to me the "war games" he played in his spare time, spreading out general staff maps all over his living room, using red, blue, and green symbols to mark off mock battles and imaginary victories. Now Häfele has taken these games one step further with a monstrous word he himself coined: *hypotheticality*. Supposedly, it lends an air of scientific credibility to the shaky conjectures he spins out in the field of nuclear energy.

Example: In a lecture to representatives of the Japanese atomic industry in 1976, Häfele was asked how one should deal with the unknown. Answer: Residual risks must be "embedded" into a matrix of other well-known risks, whether natural or man-made. In this way they will be made to appear a normal consequence of the new technology—not basically different from, say, an electrical power failure in the event of a short circuit.

But an atomic disaster, such as a reactor gone out of control, can in no way be compared to an explosion of a gas tank or a dam break. To date, all technological failures have had a tendency to heal over, no matter how deep the wounds. That will not be the case in a nuclear disaster. There is, in fact, a basic difference between nuclear destruction and other industrial accidents, one that puts it into a totally different category—its long-range consequences. Radiation escaping from a nuclear facility will continue to be devastating to humanity and the environment for an extremely long time; it is irresponsible and misleading to keep forgetting or suppressing this crucial fact. This disparagement of—indeed, insult to—the "future dimension" is extremely dangerous in any objective assessment of nuclear energy and its hazards. It makes light of the enormous consequences of the new technology by making it appear no more harmful than other advances made by mankind during the past few centuries.

Thus the dangers of nuclear development are disguised to the public. The complex, far-reaching consequences of radiation damage are buried in a mass of statistics, probability curves, and mathematical formulas to make it appear that guesswork is synonymous with solid scientific evidence.

An example of this can be found in the decisions of the International Commission on Radiological Protection. The limits they set for the radiation that nuclear facilities may release into the environment are not based on levels proven to be harmless, but on a provisional compromise between unknown risks and the utility of the moment. Even today concerned scientists claim that these officially defined norms have underestimated the hazards by permitting ten times more radiation than would otherwise be considered safe. Prof. Karl Z. Morgan, former chairman of the International Commission, admitted in the September 1978 issue of the "Bulletin of the Atomic Scientists" (Chicago): "The cancer risk from exposure or ionizing radiation is much greater than was thought to be the case some years ago."

A classic example of such an "objective" study—one that is in fact shot through with the preconceived notions and values of the team that put it together—is the Reactor Safety Study drawn up by the United States Atomic Energy Commission (AEC). Widely known as the Rasmussen Report, after Professor Norman Rasmussen of the Massachusetts Institute of Technology who headed the project, it is widely acclaimed by supporters of atomic energy as the standard work in the field. Not surprisingly, its estimates of the likelihood of reactor accidents and their impact are quite low. Professor Häfele cites it as a prime example of a "normative study of accidents."

The Reactor Safety Study, however, has drawn fire from experts. The American Physical Society has condemned the report for its faulty calculations and false assumptions, despite its broad scope and high cost. In fact, the history of this study places it at the center of a massive campaign to assuage public opinion. It was sponsored by the AEC in 1972 because its repeated assurances that nuclear accidents were "highly unlikely" were being met with increased public skepticism. In large measure this was due to the Freedom of Information Act, which brought to light innumerable internal documents by concerned scientists. No matter how they tried to rationalize it, the AEC was confronted with a credibility gap.

Among the papers requested by the Union of Concerned Scientists were:

1. A 1964–65 revision of a 1957 study made by the Brookhaven Laboratory in the hope that refined research methods might improve the pessimistic prediction of accidents made by the latter. The revision largely confirmed the previous judgment and concluded ". . . we have found in our present study nothing inherent in reactors or in safeguard studies as they now have been developed, which guarantees either that major reactor accidents will not occur or that protective safeguard systems will not fail. Should such accidents occur and protective systems fail, very large [*sic*] damages could result."

Unlike the first study, this 1964–65 revision was not made public. For a time even the fact of its existence was disputed; a reaction typical for a totalitarian government, but new and highly unusual for a democracy.

2. A third exhaustive study made from 1965 to 1971 on faulty emergency core cooling systems (ECCS) reached

some alarming conclusions. The Idaho Reactor Testing Station claimed it was "beyond the present capability of engineering science" to predict how well these emergency cooling systems would work in the event of a crisis. Nevertheless, on June 29, 1971 the AEC delivered itself of a policy statement on ECCS that declared "the industry's analysis of ECCS effectiveness to be adequate."

To reassure public opinion regarding the cooling system, the AEC sponsored an open hearing. But none of the scientists who had expressed skepticism were invited. Only "friendly" witnesses appeared, each of whom was given a numbered instruction sheet to ensure favorable testimony. Point 10 of this subsequently published set of marching orders read: "Under no circumstance contradict official policy."

In 1972 an entirely new "independent study" entitled *Reactor Safety Study (RSS)* was launched to allay all remaining doubts and it was this AEC "study" which has become known as the Rasmussen Report. That this was yet another cover-up was later established by the Union of Concerned Scientists. For example, they came across letters from the AEC to participating scientists stressing that "facts which do not support our previously established conclusions" should not even be examined. Likewise, investigations related to the safety measures in reactor construction should not be undertaken, because it was not certain whether the results would bolster confidence.

The very selection of Professor Norman Rasmussen to head the study should have been a clear signal that the initiators of the study were not interested in an objective review. Although he was not an expert in the specialized field of reactor research, Rasmussen came well recommended to the AEC, having established himself years earlier as a well-paid consultant to the nuclear industry. Somehow these activities were not mentioned in the biography of Rasmussen that was released to the press.

The data that Rasmussen and his colleagues were called upon to review relied, as they would have to admit, almost totally on material supplied to them by companies that built and operated reactors. So, right from the start, in this supposedly *independent* study they obviously served the interests of the atomic industry.*

3

Today it is generally recognized that scientists are neither independent nor free from outside influences. Yet supporters of the nuclear power industry still persist in the myth of the "objective expert" making a case on the basis of his own knowledge and a sense of conscience. Ordinary citizens are led to believe that risky projects have been carefully examined in detail by men whose sense of responsibility is unassailable. In an earlier day rulers justified their actions and misdeeds by calling upon the grace of God; today they invoke the grace of experts.

---

*The Nuclear Regulatory Commission in January 1979 officially repudiated the findings of the Rasmussen report, specifically the Report's claim that the risk of a serious nuclear accident was minimal. Many responsible scientists immediately aligned themselves with the NCR's new position, but Rasmussen himself, as dogmatic as a religious believer, insisted he had been right in the past, continued to be right now, and would remain right in the future.

All too often wishful thinking prompts the experts to pass off as certain what is doubtful at best. It is here that we may locate some of the gravest risks we face in making responsible judgments of the danger of nuclear expansion. Nor are financial gain or job dependence necessarily the reasons why the experts delude us and themselves. There are many personal motives, peculiar to the history of research in nuclear technology, which help explain why well-intentioned men of integrity become spokesmen for a course they themselves know to be disastrous, "hoping against hope" that it won't be.

I have long known three physicists working in the United States who fall into this category: Hans Bethe, Alvin M. Weinberg, and Victor F. Weisskopf. They gave their best ideas and the best years of their lives to the development of the first atomic bomb. Somehow the burden of that responsibility, the guilt they bear, has seemed to make them want to make up for it to the world by stressing the future benefits of nuclear energy. How could they possibly give an objective view of the dilemma? These scientists are caught in a tragic trap: the hopes, the work of a lifetime, the esteem they crave —all depend ultimately on their winning the nuclear gamble.

It was Dr. Weinberg who brought this home to me. For many years the head of the atomic laboratory at Oak Ridge, he became one of the most enthusiastic postwar advocates of the rapid development of nuclear energy for peaceful uses. At the same time, though, he saw sooner than most of his colleagues the hazards of the atom in big industry as opposed to the atom in the laboratory. He was the first to speak of the "Faustian bargain" that atomic scientists were offering mankind. The price we would have to pay for an "almost inexhaustible supply of energy" would be an unheard-of regimentation of our social institutions for centuries to come. Only in this way could the hazards of this dangerous "gift" be held in check.

Inevitably, Weinberg sees himself as Mephistopheles tempting mankind. But he is not comfortable in the role. During a discussion at Laxenburg in 1973 he recalled a version of Goethe's drama in which Faust made his pact with God rather than the devil. Perhaps Weinberg was making light of the matter, but the remark betrays a curious twist in the thinking of leading scientists. Although most will never admit it publicly, they are subconsciously obsessed with the notion of being forced to play God.

I once discussed the subject with Stanley Ulam, a brilliant Polish mathematician whose original application of the so-called "Monte Carlo theory" at Los Alamos after World War II provided the theoretical basis for the hydrogen bomb. In his autobiography Ulam attempted to describe the feeling of playing God that gripped many atomic scientists. He said that when scientists saw that the weapons they invented could destroy mankind, they were overwhelmed by the impact of their work: "It simply went to their heads."

The development of the atomic bomb and the V-2 rockets in World War II were conceived not to advance scientific knowledge in its own right, but for specific objectives of destruction. The pressures these developments placed on scientists changed the whole nature of scientific inquiry. No longer do researchers publicize their work only when they have results to announce. Today the stated objective precedes research. The huge sums of money involved demand that first a committee, usually from some large institution, be convinced that the proposed project is workable. The promoter of the project is under intense pressure to guarantee success. It becomes difficult, if not impossible, to admit to false leads or mistakes. In the past, trial and error made their own contributions to scientific progress. Today's scientists must orchestrate speculation with optimistic projections to keep happy a gallery of managers and officials, and allay public opinion. First impressions must be sustained no matter how slender the thread by which they are attached. Wernher von Braun was a past master at this art of "project swinging." He was lucky—his projects all turned out as predicted.

"Projectitis is an occupational hazard in nuclear research," I was told by a person intimately familiar with the field. He deplored the fact that, all too often, expensive projects were launched on weak assumptions that sold easily to those who had the money to back them. When the results were in, they frequently belied the initial optimism.

"But did any of those who were involved admit to failure?" I asked.

"None dared," he told me. "For to do so would have cut off the flow of funds. No one wanted to have colleagues accuse them of fouling their own nest."

Thus work goes on in projects that everyone knows cannot and will not work. In the end many a project winds up as a "living corpse." Surprise, surprise—another great idea that did not live up to the enormous ballyhoo that originally touted it.

I asked for a specific example.

"That's easy," he replied. "The German fast breeder project at Kalkar. It's a bottomless trough. It's been 'improved' a dozen times over, until it's hard now to find a single component that fits in with any other."

I asked how safe the Kalkar plant is.

"I'd rather not get into that," he answered. "In this line things are as crazy as in the drug and movie industries. Incidentally, that's not my comparison, but Marchetti's."

Who is Marchetti?

"He's Häfele's closest associate at Laxenburg."

There is no doubt about it, Cesare Marchetti, a member of the Laxenburg staff for energy problems, could hold his own with any Hollywood script writer. Most noteworthy is his idea of reducing the risk of nuclear reactors by "hiding" them from public view. Out of sight, out of mind—so all fast breeders, reprocessing plants, and nuclear waste disposal dumps are to be removed from populated areas and set upon the high seas. Marchetti has even found the site for the first such international energy center to be built around the year 2000—the Pacific island of Canton just south of the equator at longitude 171°W.

Marchetti's plans call for at least five huge ships, 250 meters (820 feet) long and 40 meters (131 feet) wide, to be anchored in the Canton lagoon and outfitted with breeder reactors and reprocessing plants supplied by the industrial world. Shafts bored several kilometers deep into the basalt and granite under the tropical sea and coral would receive wastes sealed in glass. Marchetti believes that these containers, heated to some 1000°C (1832°F) at a depth of 2½ kilometers (one-and-a-half miles), would make their own way down to a depth of five kilometers (three miles). The superheated rock would cool after the radioactive wastes had passed through, sealing them in for all eternity.

Rather than producing electricity, the Canton complex

would manufacture hydrogen through the hydrolysis of ocean water, a process that is still experimental at present but should most likely be operational by the twenty-first century. Huge 300,000-ton tankers would then transport the liquified hydrogen to the industrialized states to be distributed through special pipelines and pumping stations. Marchetti's boss, Professor Häfele, described the scheme as "an oilfield that will never run dry."

The nuclear community is captivated by the idea of concentrating reactors in "nuclear parks" to resolve problems of public acceptance, safety, and transportation. Remote islands have the added advantage of safeguarding nuclear fuels against accidents and potential attack during transport from one facility to another. The entire cycle—enrichment, reprocessing, and re-use, followed by disposal—could be sheltered in huge atomic fortresses.

However, such nuclear parks raise some provocative questions. What would be the effect on climate of the heat radiation released at these sites? What if an accident should occur? Would it not be infinitely more devastating than one at a single nuclear plant? Would workers—about 1,000 for each island—be willing to sign up for such dangerous labor in utter isolation? How many of them would crack under the strain? And what about the political implications of supplying most of the world's energy needs from five to ten such sites? Would they be run by multinational corporations, as Marchetti suggests, or by international organizations? Who would resolve potential conflicts?

Such questions were not spun out by some science-fiction writer. They figure in the "scenarios" that are under serious consideration at think tanks around the world—Laxenburg,

Palo Alto, Erlangen, Moscow. These reports make frequent use of weasel words that spell the difference between certainty and uncertainty—words like *soon, almost, in preparation, nearly, unlikely, except for a small percentage, presumably*. Think of all the *if*'s balanced by so many *but*'s . . . what a collection of rickety structures buffeted by the winds of shifting hypotheses!

4

One of these atomic scientists merrily spinning the roulette wheel of man's future has reminded us that the human factor is ultimately just as important as all of the fascinating formulas and exciting new technology. Alvin M. Weinberg, former head of the Oak Ridge Laboratory, is unique in the club of atomic scientists. Now in his sixties, he heads an institute that is concerned with projecting a workable nuclear future. Seldom have I seen an audience so upset as the one that heard Weinberg's keynote speech at the twentieth anniversary of the International Atomic Energy Agency in May 1977. The assembled colleagues had gathered to hear a glowing vision of a nuclear world. Instead they heard one of their own make some bleak predictions that shook them to the core.

Weinberg told his fellow scientists that they had underestimated their success to date by considering the development of nuclear power as an isolated phenomenon of modest proportions. Instead, it was likely to be man's most important source of energy. By the year 2050 Weinberg predicted that three-fourths of all the world's energy would be derived from fast breeders. Presumably by then the world would

have some 5,000 reactors each developing 5,000 megawatts
—or nine times the energy production achieved today.

So far so good. But then came the big *IF*. If we accepted
the Rasmussen Report's probability of a major accident (the
melting down of a reactor core) every 20,000 years for each
reactor, then we could also expect a catastrophe *every four
years* with 5,000 reactors in operation.

This grim prognosis set off an audible shock wave
throughout the audience. Weinberg quickly tried to reassure
his listeners that while a melt-down was serious, it likely
would cause little damage outside the facility itself. Further,
it was "fair" to assume that the probability of such accidents
would diminish over a period of time. Weinberg then offered
this cynical capstone: By that time the public would no longer care as much, having learned to live with radioactivity as
one of life's normal risks.

One can sympathize with Weinberg's attempt to draw
back from the brink of his gloomy prognosis. It is not easy to
demolish the cherished hopes of colleagues. But anyone familiar with the subject will dispute his reasoning that nuclear plants will get safer as we gain in experience. The evidence
at hand suggests the opposite. The move from laboratory-
scale models to full-scale plants will inevitably mean some
breakdowns. Yet even the number of unexpected "incidents"
has been higher than expected. Research has shown that
high-speed neutrons used in the fission process tend to pit
and scar metal parts. The results have been loosened joints,
cracked pipes, and metal fatigue.

What to do with welds that are coming apart on highly
radioactive equipment? Seal them with spot welds? Such a
procedure is hardly indicative of sophisticated advanced

technology. Evidence of the true state of the art can be gar-
nered at technical discussions by experts, such as The International Conference of Fast Breeder Technology held in
Chicago in late 1976. Of the more than two hundred papers
submitted, more than two-thirds dealt with the problem of
accidents. After the annual meeting of the German Institute
for Reactor Safety in Cologne in 1975, the journal *Atom-
wirtschaft (Nuclear Economics)* notes: "The weakest links in the
system are conventional components, particularly turbines,
pumps, and steam generators. Other critical areas include
reactor pressure units, monitoring sensors, and control
units. . . . A second category of damage can be included under the general heading of 'fires.' Within the last decade a
series of fires has crippled reactors in a number of plants
throughout the western world."

The first version of the Rasmussen Report dismissed the
hazard of cable fires that figured so prominently in the discussions at the 1976 Chicago conference. However, soon after the report was published a fire caused by a repairman
broke out in the cable distributor of the nuclear reactor at
Browns Ferry, Alabama. No one had considered such a risk.
As a result, the probability curve for accidents was raised by
a fifth, even though that too did not come close to the potential threat. Michael Grupp of the University of Grenoble
concluded that reactors studied by the Rasmussen experts
involved risks "three to eight times greater" than the overall
calculations—precisely because of the risk of cable fires.

Of course, Alvin Weinberg knows about these un-
publicized skeletons in the nuclear closet. That is why he
warned in his speech: "It would appear that the future of our
enterprise depends upon our developing a nuclear system

that fully faces up to the serious problems of core melt-down and proliferation." He emphasized which technological and institutional "fixes" were needed to make nuclear energy acceptable in the future. His special contribution to the subject of accident prevention was his call for a sort of nuclear "priesthood"—carefully chosen members who, through the centuries, would bear the responsibility for this most dangerous of all technologies. This carefully selected "order" would be given extraordinary powers to ensure nuclear safety.

5

To me, this seems the most frightening of all the risks the nuclear establishment is willing to take in its big gamble. The nuclear elite seems fully prepared to sacrifice democracy on the altar of technological imperatives. Weinberg would institutionalize an already unjust and unequal power structure. Implicit in his plan is the creation of a new type of man or woman who would function as tirelessly, insensitively, and automatically as a cog in a machine.

The idea is not new. Many scientists have held such views for some time. Having stripped away the secrets of non-human nature, science proceeded to explore and control the innermost nature of man and his society. The techniques of production, the fruits of the natural sciences, had been a way of gaining mastery over the material world. Psychotechniques and sociotechniques, the fruits of the humanities, are put at the service of attempts to control and shape human nature to the will of the politically ambitious.

Nuclear power could well provide the interface where the control of natural forces and mankind converge. Having solved only a fraction of the problems they face, the atomic engineers and managers confidently predict they will master all of them eventually, and this with a degree of almost total perfection. But even assuming that one day all the machinery (or almost all of it) will at last work perfectly (or almost perfectly), there will still remain one last incalculable, unknown factor in the equation of the atomic planners—the human factor. Will they ever be able to exercise total control over *homo sapiens?* Not unless they devise some way of breaking in and taming this obstreperous creature that has always striven for freedom and autonomy, turning him into a *homo atomicus*—a perfectly predictable, totally controlled robot.

I suspect it is just such a grim prospect that caused the young psychologist Philip D. Pahner to part company with his colleagues—and disappear. He wants no part of such a perversion of science. His former colleagues have gotten even by labeling him a spoilsport; or, as Alvin Weinberg put it to me, "He's not quite right."

What a compliment, considering the source!

THREE

Homo Atomicus

1

Sealed in zinc coffins, the bodies of Otto Huber and Josef Ziegelmüller were buried in the Bavarian village of Lauingen on November 25, 1975—the first victims of a fatal accident at a West German nuclear plant. Relatives were there, and also strangers dressed in clothing rarely seen at country funerals. Joining staff members of the Gundremmingen nuclear power station were reporters and Bavarian government officials.

Huber, 34, and Ziegelmüller, 45, had been working at the plant for ten years. On the morning of November 19 they had passed through an airlock into the reactor building to repair a leaky valve. To reach the leak they climbed down a ladder into the cramped, low-ceilinged pump room, accompanied by the radiation safety man who tested the pipes with his Geiger counter for radioactivity.

At 10:30 A.M. just before climbing in, Huber telephoned the control room, which assured him that the valve was closed. Nevertheless, he was instructed to tighten it by hand just to make sure.

Twelve minutes later, the safety man heard a muffled explosion. A cloud of hot steam billowed up through the entry hatch. Seconds later Ziegelmüller's head appeared at the opening, but whatever he was trying to say was drowned out by the hissing steam.

Rushed by helicopter to a special burn unit in Ludwigshafen, Josef Ziegelmüller died the next morning. His friend Otto Huber had been killed instantly by the steam —intensely hot at 285°C (545°F) and mildly radioactive.

Because of the possibility of radioactive contamination, both corpses were carefully checked in an isolated wing of a Munich hospital before an autopsy was performed. The official report listed negligible levels of radiation in the bodies, making it possible to release them for burial a few days later. Couched in all of the bureaucratic language one can sense official relief that the safety measures built into the plant had thwarted a major catastrophe.

Nevertheless, the deaths triggered an investigation that went all the way to the West German parliament. According to the minister of the interior the accident "could have been avoided if safety regulations had been observed." It appeared that the two men, contrary to the instructions of their immediate supervisor, had removed the packing around the valve instead of merely loosening it. They had also ignored his admonition to check the gauge to see whether water was still escaping and to inform the control room before beginning work.

Unfortunately, the gauge was not in the room where the repairs were required, but one flight below. Apparently the workmen didn't feel it necessary to take the extra steps that might have prevented the accident. Further, the safety man should have insisted that they do so. Finally, there was the question of why the system was still under pressure when such repair work was scheduled. While an official of the Bavarian environmental ministry regretted the delayed shutdown, he offered the lame excuse that "it took days to reduce steam pressure." In the present instance it had only been turned off the morning of the accident.

The Institute for Reactor Safety at Cologne estimates 25 to 50 breakdowns annually at each reactor. It regards this as a conservative estimate for recently commissioned facilities. The federal ministry of the interior maintains that all breakdowns thus far can be laid to "human error." The Institute documents such a record with a litany of "faulty calibration, faulty adjustments, faulty repairs, insufficient waiting periods, incomplete testing, etc." To err is human—an obvious comment that affords scant comfort when we consider an industry so potentially dangerous to mankind.

2

Garrett Hardin, professor of human ecology at the University of California, has cited the fallibility factor as decisive in weighing risks in industry, and particularly the nuclear industry. For at no point are human beings *not* involved in the process.

Witness: *Men* mine uranium,
       Men transport the ore to an enrichment plant,
       Men process the fuel and
       Men transport the uranium concentrates to the
          manufacturer.
       Men make reactor fuel rods,
       Men transport the rods to the reactor,

Men service and control the reactor,
Men remove the spent fuel,
Men transport it to the reprocessing plants and
Men reprocess the fuel.
Men put this new fuel to use again,
Men cope with radioactive waste,
Men get rid of it by burying it or sinking it and
Men must stand guard over the waste for
many, many years to come.

All of these steps involve fallible men prone to making mistakes. From experience we know that they will continue making a predictable percentage of mistakes in the future. Even airline pilots, whom Alvin Weinberg suggests as prototypes for the nuclear technical elite, occasionally err. Pilot error caused the collision of two jumbo jets at Tenerife; almost 600 people were killed. In an atomic disaster the victims could number in the thousands and the long-term effects might last for centuries.

Somehow, the atomic industry must beat the odds. It must cut back the accident rate at nuclear facilities to zero, an all but impossible task given the ever-present specters of fatigue, negligence, or simple confusion due to unfamiliar surroundings.

The National Aeronautics and Space Administration (NASA) has developed a first-rate system for minimizing human error. The nuclear energy community likes to point to NASA's program as a prototype for a similar effort on its part. Yet even NASA's best efforts could not prevent near disaster for the Apollo 13 capsule because someone had slipped up on a last-minute check before lift-off. Then there was the tragic fire that killed three astronauts during a training exercise at Cape Kennedy in a 204 capsule. An investigation uncovered a long list of human failings—in constructing the capsule, in lining it with combustible plastic material, in slipshod cable installations, and a dozen other such inadequacies.

Before being hired by NASA, every individual is subjected to a battery of tests and a background check. Yet a congressional investigation disclosed human error in virtually every phase of the Apollo operation, from earliest planning stages down to final tests before lift-off. Mistakes were often caused either by stress or a simple failure to follow the rules. Ironically, in its effort to design a fail-safe system, NASA had so compartmentalized jobs that those in different areas were not permitted or able to communicate with each other, thus compounding the risks.

Despite this less than perfect record, optimists in nuclear energy still believe that specially selected workers, properly trained and highly motivated, can deliver the industry into the hands of *man the infallible*. As it is, human beings remain essential in the technical system. They are known as *liveware,* to distinguish them from *hardware* (the equipment) and *software* (the program itself).

A German study of the human factor in nuclear power stations sought to devise "unequivocal character profiles" to assist in choosing the most reliable "liveware." The objective is to create an elite not only highly trained for its task, but also "in the right physical and, in particular, mental condition to use its skills to best advantage." It is deemed particularly important that the "personnel not lose their heads in event of a breakdown."

3

The nuclear industry carefully keeps under wraps which of the numerous personnel testing methods it employs. Even though many nations deny it, the fact remains that most background checks inevitably involve the police. The British have been most candid in confronting this problem. In their study, *Nuclear Prospects,* Michael Flood and Robin Grove-White point out that "all professional staff (and numerous industrial staff) at the AEA (Atomic Energy Authority) are 'positively vetted' (PV'd) prior to appointment—a procedure which entails rigorous investigation of the personal lives of individuals by security service officers, including scrutiny and a five-year review of their political associations."

The Americans are even more rigorous in their screening of nuclear industry employees in the private sector. In 1974 Congress revised the Atomic Energy Act "to authorize the Nuclear Regulatory Commission to establish employee screening programs for private companies having access to special nuclear materials." Regulations laid down in 1977 specified two types of clearances, depending on the sensitivity of the job: NRC-R, for workers allowed "unescorted access to protected areas"; and NRC-U, for workers allowed "unescorted access to special nuclear materials (SNM)"— that is, to areas where the likelihood exists that SNM could be stolen or sabotaged. This special clearance, corresponding to the military "Q" clearance, was also mandated for "drivers of motor vehicles, pilots of aircraft, as well as escorts who transport SNM by road, rail, air, or sea."

These regulations, to be applied no later than the summer of 1978, were met with a storm of protest by both workers and management. Workers worried about the time lost in obtaining clearance—weeks for NRC-R and months for NRC-U—while management worried about the high cost of the screening process. Both complained that security is already tight, what with "pat-down searches" the rule for nuclear workers since 1974. A compromise was reached in that the embarrassing search procedure would only be employed against a limited number of workers, and only under unusual circumstances. Workers also chafed under the so-called "two-man rule" that required employees in sensitive areas to work in pairs, presumably so that each would guard the other.

Donald Knuth, president of KMC, Inc., and formerly with the Nuclear Regulatory Commission, considers the "buddy system" both unnecessary and "over-zealous." Jay M. Pilant, a key employee of the Nebraska Power District, complains that no other workers are subjected to such procedures: "I don't quite understand why the nuclear industry has been singled out. It is very hard to explain to people who have been working with the company for ten, fifteen, or twenty years why they have to be watched every time they turn around. If you had workers you wouldn't trust, you wouldn't trust them any more in pairs than alone. . . ."

In West Germany, where over-enthusiasm when it comes to security is a tradition, such stringent measures are already employed in certain industries that have no immediate connection with nuclear power, but serve only to supply parts and build facilities. An official in a West German power plant commented at a meeting of the Institute for Reactor Safety in 1975: "In the future those who run the nuclear power stations must be prepared for personnel problems.

The pool of available applicants will start shrinking. When clearing someone, the police are interested in a background check that goes back five or ten years. Almost all foreign workers in the Federal Republic will be excluded by that measure."

An ever broader net is being cast by these police investigations that will soon include all who are even remotely connected with the nuclear industry. That means thousands of workers today and, with expanding atomic power facilities, the number could later on run into the millions. These will include not only military personnel and bureaucrats, but ordinary employees and workers. All must be prepared to have their political beliefs filtered through a fine mesh net.

Already construction workers at nuclear facilities are subject to background checks. This has been the case ever since the police were tipped off that terrorists had planted a time bomb within the walls of a reactor under construction. The entire structure had to be torn down and the floor ripped up with jack hammers. In the end, it turned out to be a hoax. As a reactor safety officer told me: "If we had checked out these workers a bit more closely before they were hired, we'd never have had this expensive mess."

How close is close? If a worker's political views are reviewed, then why not make doubly sure he is reliable by checking to see if he has a "normal" life-style and no personality flaws? In the United States reaction to such snooping began to make some inroads during the early seventies. But increasing unemployment has once again made it possible for employers to insist on these demeaning tests. They justify them because of the high degree of responsibility the new employee will bear in the event of a crisis.

A report in *Nucleonic News* (April 29, 1976) reveals that the United States is at least concerned with the consequences of such investigations:

> NRC staffers are now working on possibly one of the touchiest things the agency has ever handled, something with considerable potential civil rights impact. Nuclear opponents see it as the sure road to "1984" with its "Big Brother" and other societal horrors that George Orwell wrote about thirty odd years ago. The problem is establishing a standard for personnel clearances or background checks of people who will be handling special nuclear materials. Hitherto the United States government has had its so-called "Q clearances" for people [having] access to classified information. But, said an NRC source, the problem with special nuclear materials (SNM) is that there is the possibility of an emotionally unstable person employed in a nuclear fuel cycle facility and with access to SNM being able to act suddenly in an aberrant way and cause considerable harm. This kind of person would be unlikely to do something irrational with classified information, but might well make a "big splash" with SNM as a means of proving something to himself and his friends, NRC reasons. The problem NRC faces is to devise a means of detecting the instability potential in a man or woman applying for a fuel cycle job without either violating civil rights by snooping or arbitrarily—and incorrectly!—branding them as unstable or inclined to violence.

There is less concern over this protection of basic human rights in Germany. Indeed, there exists a pattern of discrimination that a survey published in the Düsseldorf *Wirtschaftswoche (Economic Week)* quite clearly revealed even when management is not concerned with such sensitive positions as those at nuclear plants. Once the investigations for nuclear plant employees or for those even peripherally involved get under way, investigations and background checks will become a way of life.

TYPES OF PERSONS	PERCEPTIONS	REACTIONS
SMOKERS	prone to illness, nervousness	watch during interview
HOMOSEXUALS	disagreeable; incapable for certain jobs, such as instructor or head of personnel	psychological tests, references, background check by an investigator
HANDICAPPED	general prejudices against handicapped; hard to fire (have they been working)	look them over, interview, check tax record
NO RELIGION "OFF-BEAT" BELIEFS	prejudice against certain faiths; consideration of regional preferences	work record, interview, psychological tests
WOMEN	not capable of leadership positions; possibility of pregnancy; general prejudices	look them over
FOREIGNERS	probably irresponsible; negative impact on customers	consider résumé and interview
LEFTISTS	probably will disrupt the work force; agitators	psychological tests, background check
CERTAIN "MAJORS" FROM THE BERLIN AND BREMEN UNIVERSITIES	could be incipient Marxists	check transcript, résumé
GRADUATES OF INTERNATIONAL MANAGEMENT SCHOOLS	overqualified for small operations	check transcript; carefully consider résumé
BACHELORS	probably unreliable	check résumé, interview
DIVORCED PERSONS	probably unreliable	check résumé

These "perceptions" set forth above are likely to be a part of the screening procedure for nuclear plant employees. In seeking the "most reliable" employees management will screen out many applicants, sometimes including those most willing to assume the risks of working in a radioactive environment. The background check and snooping soon become a way of life; once the accepted applicant becomes an employee he knows that he must now watch his every move and that of his colleagues.

Inevitably, the type of person ideal for such work is one prone to following orders. It is this predisposition toward unquestioning obedience that will mark the way he will interact with his environment, the fragile system made up of man and machine. In other situtions he might eventually come to resent the controls imposed on him as intolerable. But now the fear that he might cause a catastrophe if he fails to follow orders to the letter becomes an overwhelming restraint which justifies blind obedience. Thus the technocrats set themselves up as solicitous executive agents for a crisis situation that they alone created.

4

Since most people are not likely to pass such rigorous scrutiny today, the nuclear industry faces a serious challenge to its ambitions for rapid expansion. The search for scarce liveware is also exacerbated by the growing public criticism of nuclear energy that has frightened off many new applicants.

"When the Swedish reactor firm ASEA placed want ads ten years ago, hundreds of inquiries came in," Professor

Hannes Alfvén told me. "Today only one or two respond." Professor Alfvén was himself a key contributor to Sweden's nuclear development—until he realized its inherent dangers and withdrew his support.

Professor K. H. Bekurts, president of the European Nuclear Society, confirmed this trend when he told me with some exasperation, "One thing the nuclear opponents have achieved: They've chilled the enthusiasm of young people in our enterprise. For example, my own son is looking for another line of work. He has no desire to be subjected to the criticism and abuse that I am."

In time, the supporters of nuclear energy hope to solve the personnel problem by using robots. Ideally, such machines would do all the work. In the United States experiments have already been made with an "electronic battlefield" designed to win future wars with computers and remote-controlled missiles. However, the results proved so negative that it would be folly to use them as a model for the nuclear industry.

Total automation in atomic plants could well cause more problems than it solves. There is the danger of creating a Frankenstein monster, full of bugs and quirks, but beyond anyone's control. Highly sensitive electronic equipment responds to stimuli that often cannot be foreseen. An investigation of an accident at Windscale in 1973 revealed that no one had paid any heed to the automatic alarm system. During the previous months it had signaled so many false alarms that everyone had learned to ignore it.

On the other hand, technicians on duty at the German nuclear facility at Brunsbüttel in June 1978 by-passed the automatic security system to prevent a costly rapid shutdown of the reactor. A similar maneuver to save money was attempted some months earlier at the most modern of German atomic plants at Neckarwestheim. The result of this false economy was damage in excess of 10 million dollars. To cover up such "tinkering" (which had been expressly prohibited by operational procedures), those responsible had doctored the computer program after the fact to throw the investigative commission looking into the incident off the track. It would seem, then, that the human factor cannot be excluded from even the most optimistic hopes for automation aspired to by the proponents of nuclear energy.

In the end, man—with all his frailties—will remain an integral part of the equation. And as present workers reach or exceed the maximum allowable dosage of radiation, an ever more frantic search for "new blood" will develop. In West Germany there was the case of inmates at a shelter for the homeless being recruited. In the United States unemployed blacks have been taken right off the street.

By the year 2100 the nuclear industry will face a crisis, having increased its capacity a thousand times over the present, but with no new "radiation fodder" to tend to its care and feeding. Will the day come when every able-bodied citizen will be drafted for a period of time just to "keep the lights burning"?

Is it likely that we will develop drugs to increase the human cell's resistance to radiation? Some say scientists are already at work on this. For now, there are internationally agreed-upon limits to the radiation dosage allowed each employee. Will plant managers simply work to get the ceiling raised—as many have already been trying to do? As it is, the ceiling is already ten times higher than that allowed for the rest of the population. In short, will the individual worker be subjected to from 30 to 100 percent more radiation than he or she can "tolerate" today—or will scientists come up with an "improved" race of human beings who can tolerate huge doses of radiation?

What about using mood-altering drugs on nuclear plant employees to blot out both the risks and the perceived dangers from the individual worker's mind? Far-fetched as it sounds, such a notion must be considered if we take into account the scoundrels who are working alongside reputable scientists in laboratories and management. The post-Watergate hearings revealed that biologists and chemists regarded as perfectly respectable had accepted huge fees from secretive U.S. agencies to work on hallucinogenic drugs and other means of mind control for political purposes. One cannot rule out the fanatics who would shape an "atomic man" for their perception of the atomic future. These are terrifying visions. They can only be kept from becoming reality if the public wakes up in time and takes measures to protect itself.

Even now, the nuclear community is seeking ways to neutralize the bad publicity that has damaged its cause. To those who run atomic installations, protecting their cause from adverse public opinion is more important than protecting the public from radiation leaks. All the conventional methods of salesmanship and motivation techniques are used to attract reliable "atomic men" to their facilities and to persuade millions of their fellow citizens to accept nuclear energy.

The captains of the nuclear industry are pursuing psycho-sociological investigations to discover why large numbers of people have failed to respond to their publicity campaigns backed by millions of dollars, pounds, francs, marks, guilders, crowns, and lire. Is there some other approach that might work better? What will succeed where conventional public relations efforts have failed?

The answer to these questions has become a matter of top priority for the nuclear community. No less than three sessions of the International Atomic Conference in 1977 were devoted to the subject of "Nuclear Power and Public Opinion." Even so, the conference organizers found themselves unable to shake off their long-held contempt for the media and a concerned public that has for so long marked the arrogance of this endeavor. For example, although public opinion was on the agenda, the public itself was excluded from the sessions. They were not even admitted to the two lectures and discussion periods directly related to the subject, each held in the largest hall available to the conference. Yet the auditorium remained half empty, with police armed with rifles keeping the people out. No man in the street, housewife, school child, or student was allowed inside.

Reporters were admitted, but were banned from speaking or even asking questions. At one point a journalist rose to correct a one-sided account of the nuclear controversy. He was quickly called to order. He kept raising his hand until the end of the session, but no one called on him. The atomic experts had come there to talk about public opinion and publicity—subjects concerning which they knew very little—but they were not about to ask for any advice, opinions, or corrections from the public itself.

Public protest, previously dismissed by experts, now looms as one of the principal worries of the nuclear industry. This, despite the vigorous propaganda undertaken by all nations with nuclear energy programs to "enlighten" their citizens. According to Klaus Barthelt, top manager of Kraftwerk-Union, West Germany's only reactor builders, thousands of printed pages have been distributed ("if laid end to end they would stretch out 1,500 kilometers, or the distance from Berlin to Rome") in response to a query as to why the industry has remained silent.

Power companies have been far from silent. They grind out their own propaganda, including contests for school children: "Imagine yourself in the future. . . . What will it be like? Will artificial suns always ensure good weather? Will pedestrians move on conveyor belts? Will we have fully automatic kitchens, electric cars, and do our homework with video tape recorders?" It is never too early to start cultivating future customers.

These propagandists say nothing about the risks and dangers of nuclear energy. People who raise questions and call for a public debate all too often are smeared by the "respectable" spokesmen of the nuclear community as sensation mongers who encourage wild emotionalism and create baseless fears.

As if to make up for past neglect, power company propaganda has been stepped up. *Electricité de France* dispatches information teams to the smallest villages, even calling people on the telephone to make its case. Employees of nuclear facilities are encouraged to spread favorable publicity by word of mouth.

In West Germany, the ministry for research and development has started a nationwide sham "dialogue with the people," attempting to answer polite questions posed to government officials, professors, and even the minister himself.

The American Nuclear Society spends millions of dollars on television spots made by top advertising agencies. To influence referendums on nuclear power plants in eight states, the United States power lobby spent some $8,000,000 in 1976.

The Austrian nuclear lobby spent 25 million schillings to influence public opinion on just one nuclear reactor referendum, which it lost in the end to a loose citizens' coalition.

Last year alone the Swiss electrical industry spent 30 million franks for similar purposes.

In Japan the nuclear power lobby has had to overcome the tragic legacy of Hiroshima and Nagasaki. At first it scheduled public debates, but called the whole thing off in favor of written questions submitted to public officials. Recalcitrant communities were brought into line with bribes of new sports facilities, hospitals, and schools to be paid for out of a special fund.

In the Philippines, peasants, fishermen, and hunters are required to watch films prepared by the national nuclear authority. The media face sanctions if they fail to cooperate with the industry's publicity campaign.

"Power for Good" (PFG), a multinational public relations effort, seeks to link the experiences and pool the resources of all these various propaganda efforts. Its first public statement, appearing simultaneously in 700 newspapers and announced in 300 radio and TV spots, warned against "the danger to the peace and stability of the world that would result from a chronic energy shortage." But it would appear that PFG will have to come up with something more imaginative. The response to its first attempt was somewhat underwhelming.

5

It is obvious that the nuclear energy lobby is escalating its bid to influence public opinion. It is investing in surveys to probe the mood of the public, seeking to find a way in which to gain acceptance for atomic power. All of the "psycho-engineers" in Vienna, Laxenburg, Paris, Tokyo, and Philadelphia are agreed on one point: In the past, the deep-seated anxieties of the public have been too lightly dismissed.

An example of the work of these hidden persuaders is offered by the Austrian Institute for the Study of Conflict. A fifteen-part survey was proposed in a confidential memorandum that fell into unauthorized hands, thereby canceling the scheme—or so we are told. The study sought to establish an *Angst Katalog* (Catalog of Fears) expressed by carefully selected groups made up of both those for and against nuclear energy. The authors proposed to plumb the psyches of their subjects in a search for "possible tactics to significantly influence such opinions. . . ."

Even though the proposal had a veneer of objectivity because it was directed at both sides of the issue, the very description of the fears reveals the bias of those who formulated the survey. Advocates of nuclear energy expansion were to be asked about their "fear of being left behind, fear of weak resolve, fear of irrationality, fear of a lack of social responsibility"—all perfectly reasonable worries that would tend to make the individual quizzed appear normal no matter what the answer. However, opponents of atomic power were to be asked such loaded questions as whether they "had fear of being a devil's disciple, fear of the supernatural, fear of their own destructive tendencies, fear of hellfire, fear of the unknown." Every one of these questions hints at the irrational; the person being asked to answer them is accused of "guilt by fantasy" regardless of the response.

The Austrian government, which had been waging a public information campaign on its own, was urged to conduct such a psychological study to help it make "significant decisions, despite their apparent unpopularity, because the consequences of postponement or evasion would involve far greater risks."

The government was also warned by the authors of the study that, without popular support, there would be trouble ahead. The country simply had to be persuaded "to make realistic adjustments to the inevitable."

6

In the Spring of 1977 Beate von Devivere resigned her position as director of a study on public attitudes toward different forms of energy production in West Germany. The study was commissioned by Bonn's ministry of the interior in July 1975, when controversy over the Wyhl nuclear power station was at its height. In a public statement Frau von Devivere justified her action because the project was designed, she said, to find ways to "split the movement by uncovering conflicts between workers and farmers, housewives and retired persons, young and old."

I asked this attractive young social scientist what prompted her to make this decision. She said it was the cynical frankness with which the project's sponsors outlined their true objectives to the experts they had hired.

"It takes no special courage," she explained, "to protest publicly against deception disguised as 'public dialogue.' Those who see the light must act accordingly. I could not have done otherwise." For doing her civic duty Frau von Devivere was promptly fired from her regular job with the Battelle Institute.

The Bonn study, like so many others of its type, seeks to find out how citizens form their opinions so that the authorities can shift them in favor of nuclear power stations. The few studies that have leaked into public view go far beyond simple propaganda or even attempts to divide the opposition. In linking opposition to nuclear power to a widespread public antipathy, governments are out to win a larger war. They want to reverse resistance to the trend of ever bigger government, which is often expressed as a distrust of technology—and nuclear technology in particular.

Laxenburg's Helga Novotny suspects "that opposition to nuclear power expresses resentment toward those who profit from 'bigness'. . . . big business, big government, big science. All seem to have joined forces to provoke this massive resistance on the part of those who feel powerless and insignificant in the face of these developments."

Henry J. Otway of the International Atomic Energy Agency in Vienna predicts that "nuclear energy will play a symbolic role in the debate over the form and direction of a technologically determined future." According to him, there is no turning back from a road beset with so many problems. Instead, he is exploring possible compromises, ways to make the risks of nuclear power more palatable to its opponents. His writings suggest that nuclear advocates would be more successful if they stopped trying to disguise the dangers of nuclear development. They might do better by persuading people to accept a nuclear future in the full knowledge of the considerable risks it entails.

Otway has discovered that people are much more willing to accept a danger to which they have given their consent, rather than one imposed without it. But the opponents will consent only if they are convinced by precise calculation of the risk factors involved that the advantages of nuclear energy in the end outweigh all its dangers. An "orderly listing" of the various options, their likely effects, and an appraisal of the social benefits balanced against the calculated risks—these would help make the nuclear case.

This new strategy was tested in May 1977 by the head of the West German Nuclear Technology Society, who commented at a meeting in Mannheim: "After people have faced the fact that our civilization involves risks, they will have to learn to assume a reasonable attitude toward these risks. They must recognize that civilization, our standard of living and quality of life are not a free trip."

Risk analysts have already calculated the "fare" that people must pay for this trip. They use sophisticated equations to determine the worth of a human life and how many lives will have to be sacrificed to achieve an energy-rich society. They ask whether the huge costs involved in ensuring reactor safety are not already too high. Wouldn't it be cheaper if the industry—with state aid, of course—diverted some of the funds expended on reactor safety measures to "appropriate" compensation for irradiated nuclear plant workers or, in the case of fatalities, to their next of kin? For those who think this way, *homo atomicus* is merely a component with a price tag— the one measure of value that really counts. By this reckoning, the deaths of Otto Huber and Josef Ziegelmüller would simply have been entered on the books as "petty cash expenses."

It is astounding that the international trade union movement, which has delivered up its workers to this new serfdom, has remained relatively silent. In fact, in some countries unions have strongly supported nuclear development in the name of technological progress and job security. But in other nations workers are finally waking up to the hazards. In Australia, unions are in the forefront of the anti-nuclear movement. The Oil, Chemical, and Atomic Workers (OCAW) Union in the United States has on numerous occasions protested the health risks imposed on its membership. In England, the Flowers Report on the use "of informers, infiltrators, wiretapping, checking of bank accounts, and the opening of mail" has aroused trade unions against what they consider an inevitable consequence of nuclear power development.

Roy Lewis, lecturer in industrial relations at the London School of Economics, reaches some gloomy conclusions about the future rights of nuclear power plant employees in his study, "Nuclear Power and Employment Rights." Lewis writes:

> Nuclear disputes may lead to civil sanctions based on criminal offences or, more directly, to criminal sanctions on either workers or managers. Such intervention would no doubt exacerbate a strike; military intervention is also a possibility. This could entail technological risks, damage industrial relations, and

would tend toward authoritarianism. . . . Nuclear employment involves major inroads on employment and trade union rights. [When] a systematic curtailment of these rights begins, it is not easy to know where it is going to end. In a free society these are matters of fundamental importance which ought to be weighed in the balance of [any] argument about the future development of the nuclear industry.

FOUR

The

Intimidated

1

Lev Kowarski is a pioneer in atomic research. Along with
Joliot-Curie, Halban, and Perrin, one of the giants of the
"French school." During World War II Kowarski and his
colleagues fled the Germans and played a leading role in
building one of Canada's first reactors. Kowarski still believes that at this particular moment in our history nuclear
power is a necessary evil. But on the basis of all he knows,
Kowarski rejects the fast breeder. Its development cannot be
justified because of the high risks and the proliferation of
plutonium, the stuff of atomic bombs.

I met this unique man in 1955 at the first international
"Atoms for Peace" conference. I saw him often thereafter
and quickly learned to appreciate his intellectual honesty
and acute sense of humor. When I first asked him about the
atomic bomb nearly a quarter of a century ago, he said:
"Yes, there's a lot I could tell you. But for the whole truth
you'd have to pay me a million dollars. If I told you all I
know, I'd never get another job."

In the summer of 1977 I met Kowarski at Gif-sur-Yvette,

an idyllic community just south of Paris. This great bear of a man was just getting over a serious illness, yet was bubbling over with wit and fascinating news. I hope that some day he writes his memoirs, for few in his profession have been involved so deeply in nuclear research or have such a keen eye for observing what it was they were doing.

Kowarski has an instinctive sense for detail, starting at the beginning when only a few scientists were pursuing atomic energy in what they thought was a noble search for truth and continuing on through the days when the game grew deadly serious and fraught with responsibility. We reminisced about that long road that led from hope for a brighter future to darkness and discord.

Kowarski's secretary, now a visitor at his home, warned him it was time to leave if he was not to miss his plane. "I hope to see you again, soon," I said.

Kowarski stood there, as if searching for the right words to conclude our conversation. I will never forget what he finally said:

"If I don't run into any trouble. After all, my viewpoints are not without their risks."

"How come? Why?" I asked.

Kowarski let his massive frame slump back into the chair; he picked up a piece of paper that lay on the table and began—patiently, as he had done on previous occasions—to work out some figures. Only this time they had nothing to do with physics, but with the amount of power that would be needed to overcome the resistance represented by Professor K., which posed a serious threat to the "force field of fast breeders." Was he joking, as was so often the case? Or was he seriously concerned? He attempted to explain: *"Je les crois capables de tout!* (They are capable of anything!) Except, fortunately for me, to translate pluses and minuses into a serious equation for a technical enterprise. That's where my real security lies."

This cryptic message would not have made such a profound impact on me had I not found a letter from Ingo Focke, an engineer in Bremen, awaiting me when I got home. Focke wrote how his car and those of several colleagues opposed to nuclear development had been sabotaged in such a way that the tampering would not have been noticed before driving off. In the case of a high school principal who was outspoken in his criticism, despite the efforts of the town government to shut him up, the automobile proved his undoing. After one of his talks he had a fatal accident on the *autobahn,* wrote Focke.

Was it only an accident? Were Focke's suspicions a bit paranoid? Focke went on to describe the case of Professor Gerhard Osius, who was driving back home with his wife after denouncing the nuclear facility at Würgassen. In making a turn on a narrow street the steering mechanism suddenly gave out. "He called me to look at the damage," Focke wrote. "I discovered that the bolt holding in the tie rod had been loosened to such an extent that it could have come out with the slightest bump. At high speed the consequences could have been fatal for Osius."

Focke, of the same family famed for its airplanes, has turned into an anti-nuclear activist. He made his move after the company for which he was working asked him to approve and certify a shipment of valves destined for the Würgassen and Obrigheim reactors. He refused, since it was "common knowledge within the company that the parts would seize and bind after a relatively short period."

Focke told me how the company tried to buy him off. "I shouldn't have to worry about the responsibility—everything was insured. After the replacement valves had been installed I would simply have to certify that now they were absolutely sound, even though they had not been tested previously." After several such incidents, Focke concluded that his place was with the resistance movement.

Somehow Focke saw a connection between his new role as an "expert for the opposition" and the failure of his car's tail-lights. He described his suspicions to me after he had participated in a demonstration at the Grohnde nuclear power station: "The car was parked well away from the crowd. Furthermore, we had just arrived while the atmosphere was still peaceful. The car was left unattended until sunset, although we did notice a number of ill-disguised plainclothesmen lurking about the parking area. Knocking out someone's tail-lights is not a particularly worthwhile objective for a saboteur, but since our 2CV was a slow car, the lack of lights could have proved serious had we been struck from behind by a fast-moving car on the *autobahn*."

2

The increasing criminalization of the nuclear energy controversy should come as no suprise. A critical view of American economic history reveals all sorts of criminal activity linked to technological advances. Why should the situation with nuclear power be any different?

Would the captains of this new industry with a potential for incredible profits and power be any less circumspect in dealing with those who oppose them than, say, the robber barons of the nineteenth century? They, too, had others do their dirty work for them and then dismissed all accusations as outrageous slander and sensationalism.

In the United States the case of Karen Silkwood also began with an auto accident that killed the twenty-eight-year-old assistant at the Cimarron plutonium plant owned by the Kerr-McGee Corporation. Her body was found a short distance from her overturned car at dusk on November 13, 1954. According to the official report she had fallen asleep at the wheel after taking a heavy dose of tranquilizers.

But the suspicion that this was not an ordinary accident on the highway between Crescent and Oklahoma City loomed large when it was reported that David Burnham, a *New York Times* reporter, and Steven Wodka, secretary of the OCAW (Oil, Chemical, and Atomic Workers Union), were waiting for Karen Silkwood not very far from where the accident happened. Their skepticism grew when it became apparent that a file on alleged safety violations at the plant collected by Karen had vanished.

From 1970 to 1974, according to Karen Silkwood's report, 87 employees—including Karen herself—had been contaminated by plutonium in 24 separate incidents. In the fall of 1974 she traveled to Washington with two colleagues to file a complaint at union headquarters about the dangerous working conditions at the Cimarron plant. She claimed that Kerr-McGee falsified lab reports and X-ray photographs of the plutonium rods destined for the Clinch River fast breeder experiment.

Karen was sent back to the plant to gather conclusive evidence to back up her charges. After several weeks she believed that she had succeeded. But the incriminating documents have vanished, even though the trooper at the scene of the accident recalls seeing papers scattered around the wreck. To this day no one knows who stole the material gathered up by the police. No one questioned the motives of the five men—presumably from the plant management and the Atomic Energy Agency—who paid a visit to the car in the garage after the accident. And no one followed up on the evidence uncovered by a collision expert hired by the union. This expert found fresh impact marks on the back of Karen's Honda—as if the car had been rammed from behind.

Even before the accident Karen had feared the worst. A week earlier, she had been stopped on leaving the plant and ordered into a decontamination sequence because her overalls registered high levels of radioactivity. For three days she was subjected to blood, urine, and stool tests and body washdowns with harsh chemicals that prevented her from sleeping. In addition, Karen was afraid of what the plutonium might do to her, having recently heard a lecture by Dean Abrahamson that plutonium within the body was 20,000 times more toxic than cobra venom.

Since the source of Karen's alleged contamination could not be found within the plant, a team of investigators was dispatched to Karen's apartment, which she shared with a woman colleague. The men in white quickly found a powerful radiation source in her kitchen, some bologna and cheese in the refrigerator that was heavily contaminated with plutonium.

Presumably, this was the source of the radiation to which Karen Silkwood had been subjected. But who would have wanted to poison her? Howard Kohn, an author who has been investigating the case for over three years, has come up with the following theory: "Someone from the company poisoned her sandwich food to scare her into ending an investigation she was conducting for the Oil, Chemical, and Atomic Workers." It is an ominous accusation; friends in Europe with whom I have discussed the case cannot believe that even the most ruthless of American companies would be capable of such an act. Yet I recall similar disbelief on the part of fellow journalists in Switzerland when I informed them toward the end of 1942 that thousands of Jews were being gassed at Auschwitz. The rabbi of the Jewish congregation in Zürich had told me about it; yet every one of my colleagues dismissed the news. "Now that's really scare propaganda! Not even the Nazis would kill defenseless women and children that way."

3

"Karen was an extraordinary person. She refused to be intimidated by the company. She said what she thought, because she was very brave. And—today we know this—we did not support her enough. But she was prepared to go on, even when others were afraid to do so."

This tribute to Karen Silkwood by an official of the OCAW gives an indication of the climate of fear that pervaded the Cimarron plant. Upon Karen's death all plutonium workers were forced to take lie detector tests. Among the questions asked:

- Are you a member of a trade union?
- Did you ever have any discussions with Karen Silkwood?

- Do you smoke marijuana or take other narcotics?
- Have you ever met with newspaper or television reporters?

Whoever refused the test or failed to pass it was fired or shifted to an undesirable job in the plant. Although this intimidation was formally condemned by the National Labor Relations Board, only one of the disciplined employees ever received any sort of compensation.

The Silkwood case is a particularly tragic example of the atmosphere that prevails in some nuclear power facilities. Many who are aware of safety violations and other irregularities are afraid to speak out. They risk their jobs and pensions; they fear harassment and even death.

Some exceptional few in the scientific community refuse to be intimidated. One is the American nuclear engineer Robert D. Pollard, whom I became first acquainted with in Salzburg in 1977. Pollard gave me a detailed account of how he had wrestled with his own convictions for months before going public with his criticism of the Nuclear Regulatory Commission, the newly created security agency for nuclear energy. The NRC had been a "spin-off" of the Atomic Energy Commission, for which Pollard had worked six years.

Our chat in an old garden of Salzburg began innocuously enough on the subject of roses. "I took up raising flowers to get my mind off what was wrong with reactors," the forty-year-old engineer told me in his flat American accent. "It was just one of a number of hobbies I took up at the time. But none of them helped. I couldn't stop thinking what would happen if one of the units that we had rushed for delivery before it was properly tested suddenly went wild. Take Indian Point No. 2, for example, near New York City. I knew it was risky and tried to warn both my colleagues and my superiors. But they quickly gave me to understand that I shouldn't be worrying about such things. I shouldn't jeopardize my career. I tried to talk to my wife and explain it to her. And to my father. At first they didn't even understand what I was talking about. But when I grew more agitated and couldn't sleep nights my wife finally said: 'Oh, all right, if you can't do it any other way.' I cast about and found the Union of Concerned Scientists, and they directed me to Mike Wallace's television show, *60 Minutes*. In February 1976 I was able to tell millions what a few in responsible positions were unwilling to hear."

"What happened then?" I asked.

"The usual smear campaign," Pollard said. "They claimed I hadn't really tried to carry my objections through channels—that I was out for publicity. And that I wasn't really quite right in the head either. Do I look that way to you?"

In point of fact, Robert Pollard looks like the model of the clean-cut American. His manner is friendly, above-board, slightly rough-hewn. This often disarmingly pleasant engineer still can't get over the fact that "competent colleagues could be such common liars."

In a report on "Obstacles to NRC Staff Communications with Top Management," Pollard showed exactly how difficult it is for employees to express their doubts and worries if their superiors don't want to hear about it. Pollard wrote:

The barriers to the expression of views which differ from the party line take two principal forms—direct orders and actions by officials through whom concerns would otherwise be passed up the line. The first category seldom is so explicit that it can be expressed as: "On this date Mr. Jones told me that I should not push my concerns any longer if I want to continue working here." The fact that such orders are implied does not lessen their impact, unless one happens to be terribly naïve. The second category is the more numerous, making it clear to staffers that no action will be taken on a complaint except against the dissenting employee. The atmosphere is of a wall of silence that precludes any possibility of complaints being carried to the commissioners.

Shortly after his public bailout Pollard was invited to testify at Congressional hearings of the Joint Committee on Atomic Energy. He described his futile efforts to carry to the top his concern over lax safety procedures that would inevitably endanger the public. Most often he was rebuffed with the rejoinder "that the issues I was raising were not in my job description." Pollard once asked a senior legal staff member familiar with the history and development of the regulations why they were "so vague that neither the staff nor the applicants understand what is required." The answer: We adopted them in that way to follow the old AEC guidelines of allowing "applicants to build whatever they wanted."

Pollard also carried his entreaties to members of Congress. When two New York representatives, Hamilton Fish and Edward Pattison, proposed H.R. 4971, The Nuclear Reappraisal Act, Pollard found himself hoping that now, at last, a bill would be enacted that would correct some of the worst practices. When he discovered that the bill had been referred to the Joint Committee on Atomic Energy, which was generally favorably inclined toward the industry, he backed off from writing each member to support the bill. Pollard told the hearing that he did so because "I had been warned by a legal adviser of the [Nuclear Regulatory] Commission that if I started writing congressmen about my own views in an effort to support legislation, the most likely effect would be that I would lose my job."

It would appear that in the western world the fear of job loss is the equivalent of the eastern world's harsher methods of dissuasion. The Joint Committee had an opportunity to learn this first-hand from three experienced engineers for General Electric, who resigned their positions because, as they put it, "We could no longer justify devoting our life's energy to the continued development and expansion of nuclear fission power, a system we believe to be so dangerous that it now threatens the very existence of life on this planet."

Time and again the discussion between committee members and the former G.E. employees—Bridenbaugh, Hubbard, and Minor—came back to the issue of being able to speak one's mind in industry. One congressman asked whether it was commonly accepted throughout industry that one has to be a "yes-man, that you not question your supervisor about safety either in design or manufacture, or you will be out of a job. Do I overstate your views?"

Bridenbaugh, who had worked twenty years for G.E., replied: "There were a lot of concerns that I did not feel free to express—or, if I did, only in a general way. And then I felt frustrated in not being able to do anything about them because of budget restraints, or because it was someone else's responsibility, or that kind of thing."

The fact that not only ordinary citizens, who could be labeled as uninformed by the nuclear industry, but also recognized experts from within the system were suddenly speaking their mind made an enormous impact on public opinion. The testimony of the renegade G.E. scientists had international repercussions, inasmuch as their consulting firm, which they opened in California, caused the Swedish government to reassess its position regarding reprocessing and the final disposition of nuclear wastes. And Bob Pollard made a similar contribution in West Germany when he spoke out against the nuclear project at Wyhl. His quiet, competent testimony at a hearing regarding the project made a visibly favorable impression on the judges.

"If we only had more experts like that on our side!" said the Freiburg attorney Siegfried de Witt, who with a colleague successfully fought the Wyhl case. De Witt told me later the problem is that the other side can pay a great deal for renowned establishment experts. "As for qualified scientists who will risk speaking out openly against nuclear power —well, you can count them on one hand."

De Witt and I were sitting at a table in Hamburg with a young marine biologist. He had testified recently about the dangers of heat and radioactive effluents that a proposed nuclear plant would pose for a nearby river. But appearing as expert for the project developers was his boss, a distinguished scientist, professor at a university, and head of an institute. Nevertheless, the hearing found the young biologist's objections more persuasive and he carried the day. "My boss will never forgive me for this," he told me. "I suppose I'll have to change my job and get out of the field.

We live in a very small scientific world. Everyone knows everyone else and I'll never be given a chance to move ahead."

Almost worse than the ostracism of those who resist is the even more frequent effort to pressure participating scientists into agreement. A typical example can be found in the aftermath of a fire causing damage in the millions at Brazil's first reactor site. As reported in *Science* (May 19, 1978):

> An official investigation conducted in secret showed an incredible degree of unpreparedness for fire-fighting at the site—fire trucks, but no water; hoses without couplings; and a record of carelessness that had given rise to 85 smaller fires in the preceding year. Moreover, the investigation found that the Brazilian nuclear agency had tried to cover up the incident and the conditions leading up to it. Engineers in charge were pressured to sign a statement that there were no problems with safety precautions at the Angra dos Reis plant. This they refused to do.

But how many are there in the world who would resist such an order? Most succumb to the pressure; and, fearful of losing their jobs, verify the opposite of what they know to be true.

Concern over such incidents prompted me to propose in my keynote speech at the International Conference on a Non-Nuclear Future, held in Salzburg in May 1977, that a fund be established to provide material and moral support for dissident scientists. There must be many scientists who want to tell what they know, but fear the disgrace and financial hardship of being fired. Pollard agreed with me that such a fund might encourage scientists, who in the past had covertly passed on information about conditions that bothered them, to finally come out in the open. He noted that after he had left his job several colleagues quietly let him know that they were on his side, even though they didn't dare say so publicly.

However, Ronald M. Flügge, of the reactor systems branch, followed Pollard in resigning from the NRC. In a letter to its head, Marcus Rowden, he described how the unwritten rules of "covering up and going along" had already escalated the risks posed to the public. In sum, Flügge accused the commission of having failed to come to grips with the problems of nuclear safety.

Flügge's letter created such a stir that there was an immediate demand for still further congressional investigations. Since the Joint Committee on Atomic Energy was known to work hand in glove with the nuclear industry, the matter was brought before the Senate Government Operations Committee. Hopes that at least twenty dissidents would participate in this hearing proved overly optimistic. Ben Rusche, head of the nuclear reactor regulation division, moved swiftly to neutralize the impact of the investigation by urging all of his colleagues to share their concerns first with an internal investigator, Thomas McTiernan, whom he himself had appointed. All of these comments would be compiled in a special report and the NRC would try to act on their recommendations. Later it became evident that the exercise was less an effort to correct deficiencies in nuclear reactor safety than it was an operation to uncover "information leaks" among the staff. Luckily, the man appointed to implement the McTiernan Report made it quite clear that he would have nothing to do with such "dirty tricks."

I learned all this from Peter Kovler, a freelance American journalist working with a congressional staff. Kovler told me that all who planned to give damaging testimony—that is, data contrary to official NRC policy—at the hearing "were intimidated and subjected to pressure by their superiors." They were given to understand that if they made public statements they would be "blacklisted," with no possibility of finding another job related to nuclear energy. For most this would mean total exclusion from the scientific work for which they had been trained.

However, despite the risks, many of those intimidated got in touch with Kovler. He said that each had "insisted on anonymity. They asked me not to probe too deeply into technical problems, for fear their superiors might guess from the subject matter which experts had been talking to me. During our talks it became clear to me that several who had written critical reports had suddenly found themselves transferred to assignments having nothing to do with the problem. The result was that the issue simply was no longer pursued, or else was passed on to a staff member who would then have to go through the long process of familiarizing himself with the project. At the same time his knowledgeable colleague, fallen from grace, might be sitting several doors down the corridor, twiddling his thumbs."

Morale within the NRC staff sank to a new low. One man, speaking anonymously with Kovler, told him: "I came here a few years ago to do useful work. Now I have the feeling that the whole outfit is corrupt. I think I'll probably wind up doing what Pollard and Flügge did. Otherwise I won't be able to look anyone in the eye. I don't want my children growing up in a world run by unscrupulous scoundrels.''

In essence, these reports coming out of Washington reminded me of my encounters with German employees at nuclear facilities. In Bergisch-Gladbach I met Hans Walter Krause, a thirty-five-year-old mechanical engineer who, alone of all 1,800 employees in the plant where he worked, had dared express his doubts about West Germany's nuclear power program, particularly the fast breeder. Krause had held on to his job only because of his position within the union. But many of his responsibilities, including the training of new workers, were taken away from him and he was subjected to a vicious smear campaign.

Krause's troubles began after the plant management, in cooperation with the union, had asked each worker to sign a petition supporting the government's nuclear policy. Coming on the heels of the demonstrations by nuclear opponents at Brokdorf and Itzehoe, the petition prompted Krause to post the following note on the bulletin board: "We should make no mistake about it. This is the first step toward thought control. For obvious reasons management has avoided such a move in the past. There are some people who, for perfectly sound reasons, will choose not to sign this petition. Are they then to be put at a disadvantage—which could include being fired?"

According to Krause's friends, he was not far off the mark in describing the atmosphere at the plant. One of them said he would be in big trouble if his boss even found out that he was attending our meeting this evening. A few days earlier he had been called on the carpet by personnel for simply attending a discussion forum on nuclear power at a local adult education school. While he had not taken any part in the criticisms leveled at the atomic industry, he was given to understand that it was not appropriate for him to attend such sessions.

Another staff member had gone to a meeting where the point was made that the Rasmussen Report had predicted the worst possible nuclear accident as occurring no more than once in a million years. The employee had then remarked that such an accident could also happen sooner, rather than later—it might occur 100,000 years from now, or it might happen tomorrow. The next thing he knew, he was reprimanded for speaking out against the company's best interests. He was threatened with a lawsuit if he persisted.

The company made it clear that in such situations its interests took precedence over the constitutional right to free speech. The company rules demanded that prior approval be obtained before oral or written public statements could be made by workers. Initially, this regulation had been designed to protect patented technical data. Now it is widely used by the nuclear industry to silence outspoken employees and to stifle public debate on nuclear safety. Its use is often justified solely to prevent workers from being "disloyal to the company."

4

The reasons for such Draconian measures are obvious. Nuclear technology, whether in Germany or the United States, is expensive and at the same time utterly dependent on public good will. Like the stock market, its credibility floats primarily on the belief that there is money to be made by investing in it. Any criticism—particularly when it comes from within and is therefore presumed to be more knowledgeable—could shake public confidence and cut off the flow of public funds.

The fast breeder is especially vulnerable to such assaults. It is by far the most speculative venture in the industry and has had its problems from the start. One of its chief supporters, Professor Wolf Häfele, had already insisted as far back as early 1969, when he was project head of the nuclear research center at Karlsruhe, that the breeder could be built "here and now." Even at that time experts pointed out that at least 28 technological problems needed solving before it could become operational.

Kurt Rudzinski, one of Germany's most respected science writers, wrote bluntly in the *Frankfurter Allgemeine* about "the Karlsruhe irrationality"—the fantasy of facts and figures trotted out by Professor Häfele, who, according to Rudzinski, "has not made a single accurate projection to date regarding the costs, the economic feasibility, or the timetable for the breeder." Yet, Rudzinski concluded, Häfele remains one of the most highly regarded advisers to the ministry of science and research.

Häfele responded to this devastating criticism not by filing suit for libel, but by demanding at a meeting of the institute's directorate that any employees who criticized the Karlsruhe project be fired. He suspected that the newspaper had been fed its information by disgruntled staff members.

Since the sixties, the Karlsruhe center has insisted that it has the right to review all scientific publications by employees. Staff members must request such permission on a form in quadruplicate; only if three of the copies are returned with "approvals" is the request granted. Denials are often made not for reasons of technical security, but because the material runs "counter to center policy." I could cite examples of the sort of items that have been censored, but was asked not to do so to protect the individuals concerned. (The only other time that I have ever agreed to such self-censorship was when faced with protecting my sources in totalitarian countries.)

Believing in the notion that "democratic rights" apply also to the scientific community, several of the most prominent Karlsruhe staff members wrote directly to a socialist deputy in the Bundestag, the lower house in the West German assembly. I gained access to this supposedly confidential memo at a French research facility where it had been making the rounds. The scientists pointed out that the regulations not only stifled reporting of successes and failures, but also cut off an exchange of ideas within the scientific community—as well as denying the public its right to be informed on what was happening in this critical area.

In 1975 I chaired a panel discussion between some parliamentary deputies and several physicists at the annual conference of Nuremberg's Physical Society. The debate concerned the relationship between the state and science. Afterwards, two scientists came up to me to express their dismay that "big science" with its bloated, bureaucratic trappings had already crippled free and unfettered research. They said the atmosphere at the Karlsruhe center sometimes resembled that of Nazi Germany, a claim that I later verified myself in the course of my research.

At the Max von Laue-Paul Langevin research center in Grenoble, France, I was shown a copy of a memo sent in January 1941 to the German officer then in charge of the Paris police by a member of the German military government. Its author was Rudolf Greifeld, who later went on to become the managing director of the Karlsruhe center. Greifeld's memo asked that Paris bars and restaurants be directed to put up signs at their doors: "NO JEWS ALLOWED." Thirty years later his racist views had remained remarkably intact. In Karlsruhe he felt that if foreigners were to be hired—I quote from a colleague's letter—they should be "blonds from Sweden, not people from the Balkans." This same colleague also told me that Greifeld had one scientist spy on another who had been causing a bit of trouble, asking the former "to jot down in a notebook what he said."

How future staff members are trained at the Karlsruhe center emerges from a report I received: "Trainees, in addition to their professional instructions, receive regular lectures from an engineer Y., whom management, i.e. the personnel department, assigned to this job. In one lecture Y. painted an SS symbol on the wall, adding that once upon a time he had been sporting this symbol on his jacket and that he had looked pretty sharp in his uniform. Y. also yelled at trainees to 'Stay back three paces when you address me!' He also forbids them to lean back in their seats, insisting they sit ramrod straight."

5

Protests by scientists at Karlsruhe fell on deaf ears; here, as at other West German nuclear centers, criticism was suppressed. Professor Dieter von Ehrenstein urged that those scientists who had been critical of decisions in the field of nuclear energy should be among the first invited to participate in public debates. He also requested that they be provided with at least the minimal basic equipment to prove or disprove their point and that neither their findings nor their opinions be muzzled.

There are two separate censorship provisions that rule publications of scientists working at the nuclear research center in Karlsruhe. For one thing, the center is associated by contract with the firm "Interatom" which enforces particularly rigid controls over the views and the actions of its personnel. For another, a new agreement signed in July 1977 between the German and the French governments' nuclear industries calling for closer cooperation between the two countries now also gives the French, who are known to be specially restrictive in controlling the flow of information, a say in what may be published. They have the right to screen all manuscripts by members of the center before they appear in print, and to veto their publication if they wish.

Dr. Leon Grünbaum, a noted French physicist, had been one of the leading critics of Karlsruhe's restrictive policies. In 1973, despite assurances to the contrary, Dr. Grünbaum's contract was not renewed. I went to visit him at his home in a Paris suburb. He told me that Franz Josef Strauss, Germany's first atomic minister, as early as January 1965 steered the entire industry his way by stacking the German Atomic Commission with individuals who had held high positions in the Third Reich. I asked him to explain what effect this had had.

"It is no accident that these men would be so interested in nuclear power," Grünbaum continued. "Early on they must have told themselves that this would be a key industry wielding far more power and influence than any other. Perhaps they are also driven by another motive—the desire to arm Germany with nuclear weapons, something that is still denied that nation today."

The Karlsruhe center has very close ties with such totalitarian states as the Republic of South Africa, Brazil, and Argentina. Through its International Bureau it supplied Pretoria with a uranium enrichment process developed at the center. It also signed a much-disputed contract with Brazil which will supply that country with enrichment plants and reprocessing units designed and built by scientists at Karlsruhe and the Hoechst Company.

In 1964 the American government was still trying to prevent Germany from developing its own reprocessing plants in an effort to stem the international tide of proliferation. But Bonn, advised by Hitler's former economic aides, never missed a beat in producing and exporting the technology for making plutonium bombs. In the debate on the nuclear nonproliferation treaty, West Germany insisted on and won its preferred method of inspection: surveillance by remote-controlled sensors, rather than direct on-site inspections. By their own admission, German negotiators wanted no "prying" into West Germany's affairs.

"Remember the subterfuge in the twenties," Grünbaum reminded me. "The Treaty of Versailles limited the German armed forces both in size and in scope of their weapons. But after the Rapallo Conference in 1922 a secret deal was signed with the Russians. Now the German military elite could be trained in Moscow. I have it on good authority that the same sort of thing has been going on for years in Argentina, Brazil, and South Africa—but this time in the field of nuclear rearmament! Perhaps that explains why Bonn has delivered facilities valued at some 10 billion dollars to a virtually bankrupt Brazil. The money will never be repaid. But perhaps there are atomic bombs to be gotten from Brazil—

sometime in the future when West Germany stands prepared
to collect its past-due bill."

I reminded Dr. Grünbaum of the Karen Silkwood case
and warned him what might happen to him if he had the
evidence to back up his charges.

"What have I got to lose?" He shrugged. "I've lost my
job in Germany and, probably as a result of outside pressure,
another one here at the French national institute. My wife
left me a few weeks ago. She can't understand why I'm so
obsessed with all of this. But I can't think of or work at anything else."

A few days ago Yves Lenoir, a young French scientist,
called me in a high state of agitation. "We must do something in a hurry for Leon. He's being followed and his mail
is being opened. He has talked too much about what he
knows, and to the wrong people. I believe they are capable of
anything!"

There it was again . . . that phrase I had heard Lev
Kowarski use only a few weeks earlier. Coming from a scientist, it would have meant something quite different in the old
days. It would have suggested a message of hope: that one
day Man would be unlocking nature's secrets. Now the
words had the ominous ring of fears and incidents that have
been stalking researchers like so many dark shadows.

FIVE

The Spreading Danger

1

The first atomic bombs over Hiroshima and Nagasaki spread a shock wave of fear throughout the world—the prospect of mass destruction on an unimaginable scale. When the United States and the Soviet Union exploded their hydrogen bombs a few years later the specter of sudden annihilation confronted all mankind.

In the long run no one can live under such a cloud. Consequently, people came to accept the theory that a "balance of terror" would prevent either of the superpowers from going to war with the other.

It is a false hope. Since 1949 the nuclear club has grown from its two original members to six. First came Great Britain; then France, China, and India. Although the latter's test explosion on May 18, 1974, in the Rajasthan desert, was the weakest, it worried experts more than all of the others. Contrary to all her promises and agreements, Indira Gandhi had used fissionable material supplied by Canada and heavy water from the United States to crash the exclusive circle of the nuclear club. If India, an underdeveloped nation, could explode a nuclear device, why not others? The nuclear stalemate was over. The world faced a new, unpredictable nuclear arms race with the danger of atomic weapons being used in regional conflicts regardless of big power politics.

"Buddha smiles." That was the ironic and at the same time blasphemous code phrase flashed from the Indian test site to the foreign ministry after its successful detonation. Another reaction was shared with me by Eugene Rabinowitch, an American Professor at the University of Illinois in Urbana. Rabinowitch, one of the earliest champions of nuclear arms control, shook his head sadly. "When I heard the news I felt like someone who'd just been told he had a malignant tumor."

At the time Rabinowitch was editor of *The Bulletin of Atomic Scientists* which has on its cover a clock with the minute hand moving toward midnight. After the Indian test the hand again was moved forward. "Now all that is left," said Rabinowitch, "is to hope that somehow the malignancy won't spread—even though you know that just isn't possible any longer."

The second atomic age had begun. The age of proliferation.

2

The White House, autumn 1953: American planners at a breakfast conference hatched a campaign to gain acceptance for nuclear power, using the slogan "Atoms for Peace." They code-named it "Operation Wheaties," after Ike's favorite breakfast food. Thus the monster of proliferation tiptoed into the world arena. Today it looms even larger and threatens to become the biggest problem that will confront politicians at the turn of the century.

"The whole thing was conceived as something of a public relations effort in the cold war," I was told by Paul Leventhal, who in 1976 conducted a comprehensive hearing for the United States Senate on the international dangers of nuclear proliferation. "After the shock of the first Soviet hydrogen bomb test, President Eisenhower wanted to go before the United Nations and make it clear that Americans, as the 'good guys,' were prepared to place their science, knowhow, and financial assistance at the disposal of all those who wanted to use nuclear energy solely for peaceful purposes."

Despite the reservations of his own advisers and some foreign leaders (especially the Russians), Eisenhower went before the General Assembly of the United Nations on December 8, 1953:

> The United States knows [Eisenhower said] that peaceful power from atomic energy is no dream of the future. That capability, already proved, is here now— today. Who can doubt, if the entire body of the world's scientists and engineers had adequate amounts of fissionable material with which to test and develop their ideas, that this capability would rapidly be transformed into universal, efficient, and economic usage? . . . [We must] hasten the day when fear of the atom will begin to disappear from the minds of the people and the governments of the East and West. . . . This greatest of destructive forces can be developed into a great boon for the benefit of all mankind.

Many feared even then that nations who professed peaceful intentions could easily misuse the enormous power placed at their disposal. These reservations were dismissed by proponents of the plan for two reasons: first, control would be exercised by an international agency; and, second, the fissionable material supplied by the United States would be unsuitable for nuclear weapons. John Foster Dulles, then secretary of state, claimed that a conversion could be made from peaceful to military uses only by a difficult and expensive technology that had not yet been perfected. He condemned as alarmist the negative reactions to the plan.

However, it developed that the capacity for control and inspection was vastly overrated, while the presumed difficulty of converting nuclear fuel to military uses was grossly underestimated. Today a recognized expert in the field, Professor Albert Wohlstetter of the University of Chicago, believes on the basis of careful analysis that—in addition to the superpowers—another 18 nations have sufficient plutonium to construct several nuclear bombs each. By 1985 some 40 countries will have the capability to construct atomic bombs. In Wohlstetter's opinion the most worrisome are the smaller powers "in possession of perhaps 15 bombs"—just enough to pressure their neighbors with a threat of atomic war.

David Rosenbaum, who in 1974 wrote one of the first reports for the AEC on the malevolent use of "special nuclear materials," relied on the benefit of hindsight to deliver the following devastating verdict against the admirers of Dwight D. Eisenhower's nuclear energy policies: "The sad truth is that we've opened a Pandora's box and there is no way to return to a world safe from nuclear weapons, even weapons in the hands of terrorist groups. 'Atoms for Peace' may turn out to be one of the stupidest ideas of our time."*

What misled many political decision-makers was the belief—now totally disproved—that reactor plutonium, the by-product of "peaceful" nuclear reactors, cannot be used for making bombs. Although a bomb fueled with reactor plutonium would not have the power of one made with enriched plutonium, it would still have the explosive force of between 10,000 to 20,000 tons of TNT, or enough to wipe out 100,000 people if exploded over the heart of a city.

This was precisely the impact American scientists had reckoned for their first atomic weapons built in the Manhattan Project, the ones unleashed with such ghastly effect over Hiroshima and Nagasaki. Larger bombs 100 to 1,000 times more powerful were not built until after World War II, making those earlier weapons seem "small" only by way of comparison.

Worse yet, in a secret explosion at their Nevada test site America proved some years ago that a nuclear bomb could be detonated with plutonium of "normal reactor quality." This shocking fact was published in September 1977 as a warning to other nations. It was the same plutonium being manufactured today at plants throughout the world in Switzerland, West Germany, Italy, Spain, Sweden, Japan, Pakistan, and Argentina. Even though the plutonium first has to be separated in processing plants, the technology is fairly simple. Recent publications of Oak Ridge Laboratory estimate that it takes only a few months to build such a simplified version of a reprocessing plant. It is certainly within reach of all the nations just listed, not one of whom is in

*Nuclear Journal Reports, March 22, 1975.

possession of atomic weapons at the moment.

The problems will intensify when the second stage of reactor technology is reached: the proliferation of fast breeders with their yield of highly potent plutonium. In public hearings on the proposed construction of a reprocessing plant at the British seaside town of Windscale, Professor Wohlstetter emphasized that these reactors would yield hundreds of kilograms of "weapons grade" plutonium-239 with a purity of 96 percent—well above the 92 percent threshold for weapons grade nuclear fuel.* Wohlstetter pointed out that in his opinion the information supplied to Chancellor Helmut Schmidt by his nuclear energy advisers was "seriously lacking" as concerned the military implications of these breeders, the disposal of radioactive wastes, and the economic feasibility of plutonium and uranium.

Wohlstetter has for years served as an expert for the U.S. government, and is presently one of President Carter's principal advisers on nuclear energy. His views must have been decisive in helping Carter reach his decision to oppose fast breeder development and construction—a move that surprised a world girding itself for nuclear expansion.

As early as 1970 I was impressed by the high regard in which Wohlstetter was held during my visits to the Rand Corporation, a U.S. Air Force-subsidized "think tank" in Santa Monica, California. He was pointed out to me as one of the most brilliant and influential minds of the group, certainly more heeded in the advice he gave than the celebrated Herman Kahn. During the sixties Wohlstetter successfully pushed for a decisive change in American foreign policy which prompted the Air Force to give up numerous overseas

*Transcript of Proceedings, No. 58, p. 37.

bases that had become strategically obsolete as well as expensive political liabilities.

As an adviser to decision-makers in the critical field of nuclear development, Wohlstetter's opinions on providing expert help to government agencies deserve special attention. "The great difficulty political leaders confront," he once stated, "is that the problems are not only complicated, but demand knowledge in fields of science as diverse as physics, chemistry, atomic technology, weapons construction, and geology. They also call for a grounding in economics, systems analysis, and management, as well as in political and military considerations. No one can be an expert in all of these areas. . . .A political decision-maker often must rely on experience and knowledge that go well beyond the judgment of experts giving advice. This is particularly true in the field of nuclear technology, since the design of atomic weapons is, after all, kept secret."

At the time he made these remarks Wohlstetter revealed the sensational fact that J. Robert Oppenheimer and his boss, Leslie Groves, had written a memorandum at the end of World War II explaining that even low grade plutonium would be sufficient to construct a simple, yet highly effective atomic bomb. Had this secret data been brought to President Eisenhower's attention in 1953 he may never have launched his "Atoms for Peace" program with all of its grave consequences.

From now on, any discussion of nuclear technology must confront the issue that there cannot be any separation of its peaceful and military implications. Recognition of this fact has come slowly—perhaps too slowly—but the weight of experience allows no other conclusion. Gradually it must sink in that every nation opting for nuclear development will be viewed with far greater suspicion than ever before. That is especially true of nations having reprocessing plants or those having a history marked by aggressiveness.

Wohlstetter has this precisely in mind when he makes pointed reference to the fact that by the end of this century Japan and Germany will, if nuclear development continues at the present rate, possess sufficient plutonium to make a thousand nuclear warheads each. The warning is clear. Both nations renounced atomic weapons in solemnly sealed international agreements. But what is to stop them from reversing their positions?

Even more ominous was Wohlstetter's testimony at the Windscale hearing, seconded by an American expert on proliferation, Thomas B. Cochran, that a nation could shroud its nuclear development with peaceful intentions until the very moment it was ready to use or threaten others with its bomb.

Experts consider Israel and South Africa to be in this category. When Prime Minister John Vorster protests that his country is unjustly accused of having atomic weapons he may be telling the truth in the narrowest sense. But it is a fact that South Africa—and doubtlessly other nations as well —already has all of the active ingredients "on the shelf," ready to be assembled within days or even hours if the need arises.

Today few are likely to be fooled by such assertions of innocence. In fact, this technological sleight of hand is called a "binary operation"—the separate storage of components for illegal weapons. At a recent international symposium chemical experts warned of the dangers of chemical warfare despite official assurances that such weapons would never again be used. As with nuclear components, supplies of ingredients can be stored separately, remaining harmless until brought together. In a crisis these stockpiles could be unleashed as devastating poisonous gases.

3

In April 1975, two years before he achieved world fame as the liberator of the Lufthansa hostages at Mogadishu, Somalia, Lieutenant Colonel Ulrich Wegner received a fan letter from the deputy military attaché of the South African embassy in Cologne. The latter thanked the head of the West German elite border patrol unit GSG 9 "for helping us move our sensitive material from Cologne to our new embassy in Bad Godesberg."

Clearly, the reference was not to rare porcelain. The letter continued: "I am happy that at least it was something of a 'training exercise' for your troops. Since I had the privilege of accompanying the unit leader, I had an overview of the entire operation, which went smoothly and was quite impressive. ... . What I saw convinced me that whoever turns to you in a crisis can count on assistance and protection by the best troops anywhere."

A copy of this confidential letter and others like it fell into the hands of the African National Congress, an organization opposed to the Republic of South Africa's racial policies. These missives reveal a long-term pattern of cooperation between West German and South African government and industry in the field of armaments, and particularly nuclear development. The alliance began in 1962 with a visit by the head of the South African Atomic Energy Board to West Germany and culminated with the delivery to South Africa of a processing plant designed by Professor E. Becker, director of the Institute for Nuclear Technology in Karlsruhe. The result was a uranium enrichment facility opened in 1975 at Pelindaba with the assistance of German researchers and technicians. A year later Prime Minister Vorster several times dropped dark hints: "We are of course interested only in the peaceful application of nuclear energy. But we have the capability to enrich uranium . . . and we have not signed the nuclear weapons nonproliferation treaty." And, since then, in an interview published by Agence France-Presse, Vorster stressed that in his talks with U.S. Vice-President Walter Mondale he had "never promised" that South Africa would not develop nuclear weapons.

Bonn's continued secrecy and denials in the face of documented evidence only made the situation worse. This led to widespread charges in 1977, based on studies by Zdenek Cervenka and Barbara Rogers, of a "nuclear conspiracy" and a "nuclear axis" between the two countries designed to ensure a supply of nuclear fuel for West Germany and the eventual possibility of its own nuclear weapons. Bonn denied everything, of course, even when confronted with evidence of the continuing exchange of high-ranking government and military officials as well as scientists between the two nations.

Especially suspicious was the visit to South Africa in 1975 of Count Otto Lambsdorff, at that time the economic policy spokesman for the German Liberal Party. It was claimed that Lambsdorff traveled to Pretoria at South African expense in February, only to urge his colleagues in the Bundestag two months later to participate in Pretoria's uranium project. Franz Josef Strauss, West Germany's first minister for atomic energy, has visited South Africa at least five times since 1971, receiving in turn visits from South African atomic energy officials in Munich.

4

The possibility that more than commercial interests are involved in exporting Germany's nuclear technology is pointed up by the fact that German participation in an Argentine reprocessing experiment has been kept secret to this day. According to expert observers, a mini-lab using the Labex Milli cell developed in Karlsruhe can process more than two pounds of nuclear fuel daily. In the distinguished British magazine *Nuclear Engineering International* (February 1976) a list of reprocessing plants and projects in the western world places the annual production of these mini-facilities at 200 kilograms. Although the reprocessing plants, like so many others, suffered from frequent breakdowns and were idled for months at a time, it can be assumed that with such plants Argentina has sufficient plutonium to manufacture at least ten to thirty nuclear bombs.

Given the keen competition of Argentina and Brazil on the South American continent, the latter nation no doubt pressed hard for Bonn to assist it in its nuclear ambitions. The result was the delivery in 1975 of a complete nuclear facility, ranging from an enrichment plant to a reactor. As Norman Gall observed in *Foreign Policy* (Summer 1976), the Indian nuclear test in May 1974 had a profound impact on both Argentina and Brazil. Table talk among the elite of both nations often turned to speculation as to which of the two would be the first to have the bomb.

Leading Brazilians such as Chancellor Gibson Barboza and Ambassador Sergio Correa da Costa have long insisted that Brazil must have a nuclear weapons capability. The nation has refused to sign the Organization of American States (OAS) treaty ban on such weapons in South America. As Murilo Mello Filho, a journalist, noted in an article he wrote for the newsweekly *Manchete* in April 1973, "The bomb is something of a military imperative for Brazil, a political necessity. . . .As things appear now, as a potential superpower our choices are neither easy nor pleasant." This was the sort of statement that triggered alarm bells in Washington. It built up a resistance to further exportation of nuclear reactors to Brazil by Westinghouse, which had already delivered the first of them. The German nuclear industry suffered no such conscience qualms and immediately moved in to close the gap.

Brazil and Germany have always worked closely together on nuclear technology. As early as 1953 Admiral Alvaro Alberto, head of the Brazilian research council, visited Germany to meet with Otto Hahn, the Nobel Prize-winning physicist who discovered nuclear fission, and with German physicists, Wilhelm Groth and Paul Harteck. (Both Groth and Harteck had offered their services to the Third Reich in 1939, describing in a letter the military applications of atomic power. During the war years they had worked on the German nuclear project, developing the gas centrifuge to separate U-235.)

Groth is reputed to have told Admiral Alberto a scant eight years after Hitler's defeat, when Germans were forbidden to develop their own nuclear technology, "Just give me the necessary means and we will develop a prototype. Then we'll all come to Brazil and set up an installation." Shortly after, a secret agreement was struck between Germany and Brazil, calling for three Brazilian chemists to go to Germany for specialized training and for the ordering of components by Brazil from 14 German companies.

Later, according to Norman Gall's article in *Foreign Policy,* Alberto told a parliamentary investigation in Brazil: "Germany is an occupied nation. It would create an international crisis if word leaked out that it was prepared to manufacture enriched uranium."

But the crisis never materialized. American authorities learned about Groth's shipment and confiscated it at the last moment at a German port. As soon as the occupation of Germany ended, cooperation with Brazil came out into the open. Beginning in 1956, Germany achieved a position second only to the United States in nuclear dealings with Brazil.

In 1975, KWU (Kraftwerkunion) contracted with Brazil to deliver, with German Goverment backing, a complete facility to the South American country. Not surprisingly, this revived the suspicion that West Germany was seeking a foothold in nuclear weapons technology in the Third World. Bonn rejected these charges as being out of hand, claiming merely that it was "seeking nuclear independence." Yet it is hard to believe that commercial considerations were the only motivation for an agreement that for the first time caused a serious rift in the cordial postwar relations between Bonn and Washington. To the rest of the world, including Germany's old friends, it seemed more like a ruthless abetment of nuclear proliferation in Latin America.

London and Washington were particularly concerned about Bonn's fast breeder project. Both nations had shelved plans to build such reactors, which were deemed a threat to world peace by President Carter. Despite these warnings, Germany pursued its breeder policy in cooperation with France and its billion-franc *"surgénérateur* project." The conservative French magazine *Le Point* viewed this joint effort of the two major continental powers as a drive toward future technological superiority. Thus, despite their competitiveness in other areas of the nuclear marketplace, there seems a determined effort by both Germany and France to surpass the United States in technology by the turn of the century.

Isn't all of this reminiscent of previous German expansionist policies? During the thirties the German people were won over to a similar policy by such slogans as *Lebensraum* and *Volk ohne Raum*—a nation without enough land for its people. Today they are frightened into believing they will soon be a *Volk ohne Strom*—a people without electricity. Yesterday the slogan was *Deutschland über alles;* today it might as well be *Plutonium über alles.*

President Carter again appealed to the assembled delegates of forty nations in Washington during the fall of 1977 for a ban on the spread of fast breeders and reprocessing plants for plutonium. He called for an internationally controlled "uranium bank" to supply the world's needs and offered the United States as a disposal site for atomic wastes as an alternative to having it reprocessed elsewhere.

Once again Germany was spokesman for those nations that wanted no restrictions on fast breeder development. According to the conservative newspaper *Die Welt,* the conference was a triumph for the German policy: "All forty nations voted to keep their options open in nuclear technology and not to restrict the development of plutonium reactors until the completion of a two-year study called for by the London economic summit."

Despite Bonn's protests that others "misunderstood" its export policies, world public opinion remains skeptical of West Germany's benign intentions. It would feel reassured only if Bonn declared a moratorium on its nuclear program and suspended its exports of nuclear technology. That would be the only way to dispel fears that a future German government under Strauss's influence might not use nuclear power for military purposes. As one American observer at the Salzburg atomic conference told me, "If the Germans continue down the road of breeder development, in the end we shall have to follow them despite our misgivings. For the moment, though, we are still trying hard to prevent a plutonium world." Jeremy Bugler, a correspondent for London's *New Statesman*, told me in the same connection that a high official of the U.S. State Department had offered to him the gloomy prediction: "The Germans may thus become guilty of the second holocaust in this century."

5

When German supporters of the *Plutonium über alles* policy are confronted with the dangers their endeavor poses for all mankind, they inevitably reply that international "safeguards" will prevent the spread of nuclear weapons. Yet, even in this effort, Bonn seems to be calling for a hands-off policy.

In their book *Nuclear Energy in Germany*, Karl Winnacker

and Karl Wirtz record the dismay of the German government early in 1967 at the United Nations' proposals to establish via a treaty an international security system to halt nuclear proliferation. One of the features of the treaty, pushed for by the United States and Russia since 1965, called for the inspection of nations as yet without atomic energy, while leaving undisturbed the members of the original "atomic club."

Germany has opposed since the beginning Washington's inspection policies. Former Chancellor Konrad Adenauer was so disturbed by them that he talked of a "second Yalta." Other German politicians warned of an impending "American-Soviet atomic conspiracy." After the first reading of the proposed treaty Karl Wirtz hurried back to Bonn to report to Chancellor Kurt Kiesinger. There he huddled with the defense cabinet—a rather curious forum for the discussion of the peaceful uses of atomic power.

As negotiations progressed, the German delegates opposed "an inspection army under the aegis of the UN or the International Atomic Energy Agency in Vienna" and the "exorbitant cost" of supporting it. They missed the point that such an expense would have been but a tiny fraction of the worldwide arms budget. More compelling was their argument that these inspections "would unlock the gates for industrial espionage." As the leading industrial nation without atomic weapons, Germany had written into the final draft of the treaty the clause that research facilities should, "whenever possible," be monitored by instruments rather than men. This would help ensure that both the number of inspectors and their rights of surveillance be held to a minimum.

Riddled with similar caveats, the final treaty in 1969 is virtually impotent. Only 95 of 155 potential signatories have ratified it, with three of the six existing atomic powers—France, China, and India—among the abstainers. Non-nuclear nations working to establish themselves in the field (e.g., Egypt, Israel, Argentina, Brazil, Pakistan, Spain, South Africa, and Taiwan) have ignored or delayed ratification "to keep their options open."

Today there are 80 inspectors working for the International Atomic Authority in Vienna, affixing their seals on nuclear facilities in nations where they are admitted as well as setting up surveillance equipment and recording the flow of fissionable material. (Only 43 inspectors were actually on the job in 1976—with the task of inspecting 400 plants spread over several continents.) These inspectors are not allowed to follow up on violations they may detect, but must report them to the appropriate agency in the country where they occurred. This is rather like telling the fox that the hen-house has been raided. Word is also relayed to Vienna, but policy forbids publication of the violations record. In short, there is no penalty involved, not even the censure of world public opinion.

From the beginning, the objectives of these "safeguard" policies were less ambitious than the public had been led to believe. It was hoped that the warnings made by inspection teams would buy a little more time. Max Iklé, disarmament adviser to former President Gerald Ford, came to the unhappy conclusion: "Safeguards are a burglar alarm, but not a lock." What needs stressing is that the alarm would sound too late, if at all, and in such a way that few would be able to hear it. There is an unreal, kid-gloves treatment that nations afford each other in the matter of safeguards—much more restrained than a state's actions toward its own citizens. But the former have power; the latter none.

For the time being, little can be done to halt the secret growth of atomic weapons capability. But there are plenty of theories, discussions, and proposals on the subject. The nuclear plague is seen to be spreading in three ways:

1. Vertical proliferation—an expanding arsenal of nuclear weapons by states that already have them.
2. Horizontal proliferation—the acquisition of atomic weapons by nations that did not have them.
3. Non-governmental proliferation—nuclear weapons in the hands of both external and internal terrorists; certainly the most ghastly and difficult to control form of proliferation. It also raises the specter of the secret support of such groups by one government against another, thereby reducing the prospect of immediate retaliation against the instigator.

The burgeoning growth of unaccounted-for supplies of enriched uranium and plutonium deeply worries such experts as Frank Barnaby, Paul Leventhal, Joseph Rotblat, and Walter C. Patterson. As a stop-gap measure, they call for stringent international control of all nuclear material with, eventually, a ban on all fast breeders in order to prevent the prospect of a world economy based on plutonium power.

At the present growth rate of atomic programs, by the year 2000 there will be enough fissionable material scattered around the world to make more than a million nuclear weapons. The larger question is whether there will be enough prudent leadership and skilled technicians to shore up international security in a world awash with poisons and destructive potential.

The despair expressed by David Lilienthal, first chairman of the U.S. Atomic Energy Commission, is understandable in the face of this looming "nuclear anarchy" and international uncertainty. In testimony before a senate committee considering the question of nuclear proliferation, Lilienthal expressed relief that he was no longer a young man—and deep concern for the fate of his grandchildren.

SIX

Atomic Terrorists

1

Today it is widely assumed that only gangsters or political criminals would use the "ultimate weapon" of nuclear terror. This is a narrow assumption that ignores man's inventive capacity for evil. What about high-principled idealists who would achieve their ends by using any means? Such a possibility, virtually ignored at the time, was sketched out in February 1969, in the relatively obscure *IR&T Nuclear Journal* (Washington, D.C.). All the more reason to recall its implications at this time, now that the danger looms ever larger:

A group of physical and social scientists from several nations organizes to enforce worldwide nuclear disarmament. The group includes nuclear scientists from the United States and the Soviet Union who are experts in nuclear weapons technology, as well as several people who have very large private sources of capital.

The group contacts a large criminal organization

127

in the United States, offers to pay $10 million for several hundred kilograms of fully enriched uranium and plutonium that are to be taken by armed theft from two facilities in the United States that have large fast critical assemblies. The secret group helps the criminal organization plan the thefts.

The nuclear weapons experts in the group design two types of nuclear explosives: an especially lightweight fission explosive, and a large thermonuclear explosive. The designs are such that the non-nuclear components can be made with materials that are easily available in any of the nuclear weapons nations, using commonly available machine tools.

After successful theft of the special nuclear materials, these are smuggled by various means into each of the nations that has nuclear weapons, using the services of an international criminal organization that is paid $20 million for the smuggling operation. The non-nuclear components of approximately five of the light fission explosives are secretly fabricated in each country, and the assembled nuclear explosives are hidden in both heavily and sparsely populated parts of each country.

Several multimegaton thermonuclear explosives are secretly constructed in the United States and hidden in seagoing fishing boats that are purchased by the group.

When all the nuclear explosives are assembled and in place, the threatening group states its demands in letters simultaneously sent to the heads of state of the nuclear weapons powers, to several prominent newspaper editors, and to the Secretary General of the United Nations. The letters demand that all the nuclear weapons states turn over their nuclear explosives and nuclear warhead delivery vehicles to a new disarmament agency to be organized immediately by the United Nations, submit to detailed inspection by the new agency, and place all civilian nuclear facilities under a system of safeguards managed by the International Atomic Energy Agency. Failure by any nation to meet the demands, according to a specified schedule, will result in the explosion of nuclear devices in cities in that nation. The letters state that the threatening organization has both low yield fission explosives and high yield thermonuclear explosives, and that a thermonuclear device will soon be exploded in a remote area of the world to demonstrate that the threat is real.

One week later, a 20-megaton explosion is detonated in one of the fishing boats in a remote area of the South Pacific. The demands of the threatening group are restated. The group is not satisfied by subsequent action by any of the nuclear weapons states within the following month, arranges for the simultaneous detonation of 5-kiloton devices in remote areas of each of the nuclear weapons states. Extremely strong international pressures of many kinds begin to force the threatened nations to meet the demands of the threatening group. Several low yield detonations in heavily populated areas are required to demonstrate that none of the nuclear powers has organized the threat.

* * *

Subsequent development of the scenario is left to the interested reader.

Its author, Dr. Theodore B. Taylor, achieved even greater clout with yet another apocalyptic vision of the future, one that continues to unnerve decision-makers in his own and other countries. It appeared in the same journal and was later widely reprinted:

> During a State of the Union message by the president to a joint session of Congress, one of a band of terrorists drives an automobile with a nuclear explosive in the trunk as close as he can to the Capitol, stalls the vehicle, sets a ten-minute time-delay fuse to detonate the bomb, and runs for cover. The explosion destroys the Capitol, killing the president, all the attending members of Congress, the Cabinet, the justices of the Supreme Court, the joint chiefs of staff, and several thousand people outside the building.

In Taylor's view, the Capitol is the most significant target for those who have at their disposal basically one "lousy fission bomb." Taylor uses the word *lousy* not to dismiss lightly the enormity of such an attack, but to emphasize how easily terrorists could avail themselves of the opportunities to make such a weapon.

As the developer of the smallest, lightest, and most powerful of the fission weapons made by the United States, Taylor brings the full weight of his experience to bear on the devastating implications of nuclear terrorism. In his view, any reasonably intelligent group of terrorists can assemble such a device from easily obtainable components in a back room.

How much easier it would be for a determined group with some technical competence. Has it happened already? What about the 100 kilograms of highly enriched uranium— "enough for several fission explosives"—that disappeared in 1965 from an American fuel fabrication plant?

Taylor's deeply felt sense of responsibility in pointing out this hazard served less to arouse the public than to grate on its collective nerves. For years his gloomy predictions were shrugged off with sympathetic smiles. I must confess that at first even I considered him something of a crackpot. Out of a clear blue sky he called me in 1967 during his stint with the International Atomic Energy Agency. I knew nothing about him, but listened closely as he explained to me during our walk through the narrow alleys of Grinzing near Vienna that I must arouse the public to the danger of homemade nuclear weapons in the hands of gangsters.

At the time we talked it was still widely believed that atomic weapons could only be built by experts at great expense in specialized laboratories on the order of a "Manhattan Project." Then, too, terrorism was not at all the problem it was going to prove in the seventies.

Taylor stressed that the nuclear bandits needed only to lay their hands on sufficient quantities of fissionable material through theft or bribery. The ever increasing stockpiles were so poorly guarded that any raid using diversionary tactics was likely to succeed. The details on how to build a bomb up to 10 kilotons had already been widely circulated by the U.S. Atomic Energy Commission. Any device even approaching, however imperfectly, these "secret" plans would work.

"Believe me, just one of those things blowing up in the middle of a city during rush hour would destroy eight city blocks, kill tens of thousands immediately or over a period of

time, and force the long-term evacuation of the entire area,"
Taylor said quietly, as if he were holding a lecture, while all
around us revelers were cheering each other on, drinking
new wine.

"What you're telling me is too farfetched. I simply can't
believe it and can't imagine anything like that ever happening," I said.

We had been sitting at a café table. All around us people
laughed and sang, evidently enjoying life. But Taylor looked
at me glumly. "Yes. But that's the way it is." His tone of
voice indicated that he had given up trying to convince me.
He raised his glass: "To your health! So long as we still have
it!"

2

Many experts still believe that Taylor's fears are exag-
gerated. They consider security at atomic facilities so tight
that no thieves or terrorists could gain entry. Further, they
persist in the belief that no group of outsiders, no matter how
expert or how much nuclear material they had on hand,
could make a nuclear device. Laymen would find it too difficult to master the technology for implosions, they insist.
But many are concerned with the implications of Taylor's
warnings.

Thus, a study entitled "Nuclear Proliferation and Safeguards" published in 1977 by the Office of Technology
Assessment (OTA), which is not given to snap judgments,
allowed:

There are probably groups at large in the world today
that possess or could acquire the resources necessary

to become nuclear adversaries, if they wanted to. That is, they might be able to sabotage a reactor, steal fissionable material, build a dispersal device, or possibly even a crude nuclear explosive device.

On March 22, 1978, Dimitri A. Rotow, a young Harvard economics student who admitted that he was "not terribly good in physics," appeared before the Senate hearing on the Omnibus Anti-Terrorism Act chaired by Senator John Glenn. A few months earlier Rotow had prepared a term paper on "Nuclear Weapons Design and Construction." Experts who read it claimed it was "the most extensive and detailed exposition of things to think about—and how to think about them—in the design of nuclear weapons" ever seen outside the field of classified documents.

The fact that an outsider, using readily available sources, could prepare such a project so alarmed Dr. Leonard Weiss of the NRC that he immediately sought out Senator Abraham Ribicoff and even notified President Carter to discuss what implications this might hold for United States policy. If nothing else, Rotow's paper confirmed Taylor's thesis, previously laughed off, that anyone could build an A-bomb.

In cooperation with Mason Willrich, an attorney and disarmament specialist, Taylor had earlier prepared a study under a Ford Foundation grant on "Nuclear Theft: Risks and Safeguards." It is no overstatement to say that it hit the scientific community like a bombshell. Fear that it might put ideas in the heads of terrorists caused the Atomic Energy Commission to put it under wraps. But bits and pieces leaked out, notably in newspaper columns by Jack Anderson. Within a year the report, with some details presumably omitted, was finally made public. Taylor, who by that time was no longer in government service, had long argued that it was much more dangerous to keep silent on the perils of atomic power than to give them the widest possible publicity.

Slowly Ted Taylor's colleagues changed their views about this "professional pessimist." By 1976 *Current Biography* noted that he was now ranked by the scientific community as one of the most imaginative and prophetic physicists of the atomic age.

It was not an easy route for Ted Taylor. Somehow he had to make credible his claim that 3 to 10 kilograms of ordinary reactor plutonium would be sufficient for persons of average technical ability to make atomic bombs. During his time at Los Alamos he had uncovered calculations which proved just that. They had been kept secret, possibly in an effort not to jeopardize the civilian nuclear program. Taylor convinced many, including AEC security chief Carl H. Builder, that there were simple "recipes" for the construction of such bombs, workable enough for anyone willing to risk failure or accident. He also showed how shockingly inadequate the security of fissionable material was in the plants and during transit. In the early seventies seizure of such material by nuclear blackmailers would have been child's play.

Taylor's revelations were backed up by an even more dramatic plea for stringent controls within the private sector of the nuclear industry made by David Rosenbaum. The nuclear establishment responded with the dispatch and adaptability that is so typically American, developing by 1973 a strategy to combat clandestine fission explosives (CFE). Defensive measures taken against potential internal enemies began to resemble those adopted toward foes abroad, replete with general staff planning, military maneuvers, and electronic surveillance—in short, a deadly serious anti-civilian war game.

The only difference was that, in this case, security was more on guard against small, determined groups than against a mass uprising. Would three to six attackers suffice in seizing special nuclear material (SNM) or would it take fifteen terrorists? What would be their tactics? What sort of weapons would they carry? Would they not have as co-conspirators "insiders" who knew exactly where to go and what to do? These and other questions were posed by the government to "think tanks" such as the Rand Corporation in Santa Monica, the BDM Corporation in Alexandria, Virginia, and a special division of its own Office of Technology Assessment.

It was quickly determined in all of these studies that technological centralization had made society vulnerable as it had never been before. A handful of revolutionaries could, in theory, cripple entire cities and states if they knew how to strike at the nerve centers of production, transportation, and energy.

In a study prepared for the Justice Department, "Terrorism and the New Technology of Destruction," R. W. Mengel documents 4,500 terrorists acts from 1965 to 1975. He presents a chamber of horrors of modern methods of sabotage, ranging from laser beams to biological warfare. But the atomic threat still looms larger than all the rest, even though botulin toxin—which is more poisonous than plutonium—might prove more deadly if poured into drinking water.

Senator John Glenn stated that no less than 44 serious

nuclear threats had been made in the United States since 1977. There are probably many more occurring elsewhere, except that usually an effort is made to cover them up. At times that isn't possible, such as the attack by Basque nationalists against the reactor at Vandellos, Spain, under construction since 1972. Two died in that incident.

In a letter to his parents in August 1945, written after he had heard over the radio news of Hiroshima, Ted Taylor expressed the hope that fear of this weapon would forever make war unthinkable. Today all of his energies are directed toward thwarting nuclear civil war. He urges security measures that will make it impossible for terrorists to seize nuclear material. But it would appear that total security is going to prove just as illusory as Ted Taylor's hope for permanent peace in 1945. Internal security will remain just as shaky as the defenses against foreign enemies. Every new defensive measure brings about some new circumvention by terrorists; harsh security methods will be met by increasingly violent or cleverer assaults. Another legacy of the atomic age.

3

Bruce Jenkins, an expert on terrorism with the Rand Corporation, told me in 1978 some of the defense problems posed by this "inner perimeter" in the United States. Indeed, he and his colleagues at the most prestigious of American "think tanks" worry more now about internal security than about the defense postures to be taken against external invaders.

Those responsible for maintaining law and order will make every effort to plan for all of the contingencies likely to arise from terrorist threats. But the plain fact is that such bureaucratic plans cannot ever hope to cope with all the desperate schemes of twisted minds. Even if 2 percent of all the funds spent on nuclear development—which would add up to several billion dollars today—were invested in security measures, as Taylor and Willrich suggest, it is still not likely that iron-clad security could be provided. The inherent risks of nuclear development are thus compounded by the incalculable hazards of terrorism, a danger that is likely to increase in times of political discord or social upheaval.

For one thing, the enemy today is less familiar than in previous encounters. Often groups will form spontaneously, their composition, objectives, and tactics unknown to national security forces. The previously cited OTA report on "Nuclear Proliferation and Safeguards" makes the point that we cannot count on being informed in advance by "adversary forces" as to their objectives:

> There may already be, or there may appear in the future, new kinds of adversaries, or special sub-classes of adversaries that have not yet been identified, who might be more likely to use nuclear means to achieve their objectives. Threats to nuclear facilities or those involving the malevolent use of nuclear material may emerge on a different organizational and mental plane. . . .It is difficult to say now, what new entities may emerge in the coming decade.

And one must add: What about the coming centuries? Is it not patent idiocy to scatter throughout the world facilities with a technology that lends itself to the most devastating weapons and which are vulnerable to social and political conflicts? Sooner or later, must not the dragon's teeth sown by this generation spring forth?

4

How easy it is to penetrate nuclear facilities has been demonstrated by the so-called "black hat" teams, the mock opposing force in the war games played at the Sandia Labs of Albuquerque, New Mexico, who regularly penetrated security to sabotage the plant or steal plutonium. Testifying before the Senate's Government Oversight Committee in 1976, Sandia's director of security Orval E. Jones said that these war games revealed considerable gaps in security.

Furthermore, it is unlikely that such mock teams, who are only playacting terrorism, could ever operate with the devious fanaticism of the real intruders. To make the exercise completely realistic, would not real criminals have to be employed? Such a suggestion was offered in *Nucleonics Week* (November 7, 1974) by Charles Yulish, head of a well-known firm of security consultants: "We want to offer 'black hats' for hire," Yulish stated, going on to mention a "convicted felon" as a key member of a team of specialists who would "pick out any loopholes in [your] security system." The article argued that "AEC 'black hat' studies of safeguards are done by 'white hats' posing as 'black hats' and thus cannot be as effective in examining facilities as men who have [actually] done breaking and entering. . . ."

A further comment that the "expert saboteurs" of the Vietnamese War should prompt the AEC to formulate a more "realistic safeguards appraisal" may have encouraged the NRC's top security expert, Manfred von Ehrenfried, to hire former Green Berets for his simulated attacks on nuclear facilities.

When I met Ehrenfried at the Hotel Madison in Washington, D.C., he was a private citizen, having been fired by the NRC for continually exposing its lax safety measures. Similar charges had also been made by James H. Conran, a respected member of the Division of Safeguards, not only in letters to his immediate superiors, but also in reports to the chairman of the House Committee on Interior and Insular Affairs and, finally, to the president of the United States. Conran noted, in part, that the "existing safeguards at the NRC licensed facilities are afflicted pervasively by serious and chronic weaknesses which pose serious potential hazards to the public health and safety, and to common defense and security."

The initially classified report of the NRC task force set up to examine Conran's charges affords an interesting glimpse of the inner workings of the nuclear security system. To one familiar with the nuances of a whitewash, it is painfully clear that neither NRC nor the other units reporting to it cooperate as closely as might be expected. Nowhere does the task force report specifically lay to rest Conran's charge "that there is a general limitation, perhaps suppression, of information." Rather, an attempt is made to show that this respected official, charged with an extremely urgent task, was in some instances denied access to necessary information for lack of a demonstrable "need-to-know." But there is no proof given that other NRC officials were provided with such information, either.

Conran claimed that data in the following three critical areas were denied him:

1. Questions of the relative ease and the likelihood of success in the design and fabrication of clandestine fission explosives (CFE Information).
2. The implications of two types of nuclear incidents:
   (a) threatened use of clandestine fission explosives and threatened or actual malevolent acts occurring at nuclear facilities (Incident Information).
   (b) serious indications of past illicit diversion of large quantities of strategic special nuclear materials (Diversion Incident Information).
3. Knowledge of people having motives or plans involving malevolent nuclear acts (Intelligence Information).

The motives for suppressing the full impact of Conran's efforts are obvious. Just as in the cases of Robert Pollard and the three General Electric engineers, data about lagging safety in nuclear facilities had to be kept from the public. The risks of nuclear accidents and the political vulnerability of the nuclear establishment simply have to be covered up at all costs—or how else will citizens be persuaded to sustain and pay for them?

A particularly weak link in the American nuclear energy system is the transportation of fissionable material by road, rail, and air. To seize it a determined gang would not have to cope with the walls, moats, fences, barbed wire, and electronic surveillance units of the nuclear facilities. Granted, special trucks known as safe security trailers are usually disguised as moving vans. They are bombproof and have smoke and gas ejector systems to ward off attack. In future they will travel only in convoy and must give a continual fix on their position to a control station. As one driver put it, "I won't even be able to take a leak by the side of the road without their knowing what I am doing!" In the event of an attack, roadblocks will be set up ahead of and behind the convoy in an effort to head off the attackers. The men chosen to drive these convoys are among the most rigorously screened of all AEC employees.

Even more macabre is the possibility of a terrorist assault on a heart attack victim with an electronic pacemaker powered by Plutonium-238, a substance 280 times more deadly than Plutonium-239. In 1977 David Krieger, a political scientist at the University of California in San Francisco, warned in an article in the *Annals of the American Academy of Political and Social Science* that the quarter of a gram of Pu-238 present in a pacemaker could contaminate 37,500 cubic meters, according to the calculations of Willrich and Taylor.

In recent years, twenty such Pu-238 pacemakers were produced monthly in the United States. Unlike battery-operated devices, pacemakers powered with plutonium will last a lifetime without subjecting the patient to an operation every few years to replace a battery pack. Krieger's article now places their production in jeopardy, since the hazards of potential terrorism might well outweigh the very real benefits of such pacemakers for victims of heart disease.

It is too naïve to expect similar forbearance among the supporters of nuclear reactors? If only they could see these facilities as the pacemakers of a faltering technology, they might then also recognize that no security plan is foolproof. Walls, fences, and guards may keep out intruders from outside, but who is to protect against the enemy within? Think of all of the impregnable fortresses throughout history that have fallen to treachery.

Thus, current war games must also take into account the probable presence of a "fifth column" in the event of an attack on a nuclear facility. Indeed, there is already one documented case of an insider escaping with nuclear material: Eliodore Pomar, director of Euratom's nuclear research institute at Ispra, who vanished in 1974 with some radioactive material that he allegedly was going to make available for a neo-fascist coup. As usual, Euratom suppressed all information concerning this theft.

The possibilities are endless. Michael Flood, a young Englishman studying the problem of atomic sabotage, explained to me that many of the guards at nuclear facilities are losers who have failed at other jobs and exhibit fascistic tendencies. It is conceivable that a group of such guards might threaten a government with blowing up a plant if the latter did not act more energetically against leftist or "subversive" elements. On the other hand, leftist governments might find themselves confronted by similar demands from the other side. Such an event could have occurred in the Soviet Union when Beria was toppled from power in 1953. As head of the security forces he could have ordered seizure of the few facilities open at that time as a last desperate move to save his own position.

Game planners have hundreds of such variations to play with, some on the mark and others no doubt too farfetched to be of real concern. But in view of the enormous dangers posed by nuclear potential in the hands of terrorists they have to be ready for any eventuality. The number of victims held hostage in such a threat could number in the thousands.

One such nightmare was vividly described to me by Dr. Bernard Feld, head of M.I.T.'s nuclear and high energy physics research unit. In it he imagines the mayor of Boston calling him into urgent consultation after receiving a threat from a terrorist gang that they have planted a nuclear device in downtown Boston. He also receives confirmation that twenty pounds of plutonium are missing from government stockpiles. The mayor shows him a crudely drawn blueprint and the terrorists' outrageous demands. Having participated in the creation of the first atomic bomb, Dr. Feld knows that the device can work, imperfectly perhaps, but with devastating consequences. What should he tell the mayor? Should he advise giving in to the blackmail? Or should he risk destruction of the city?

An official report on "Disorders and Terrorism" published in Washington in 1976 deals with the credibility of threats employing nuclear weapons or biochemical agents. Despite the science-fiction quality of many such claims, the report warns that the dangers might be very real. Any response by the police must be based on the likelihood that the destruction threatened could very well be unleashed if push came to shove.

5

Contingency plans for terrorist assaults in an atomic state call for a mobilization of police and armed forces on a scale hitherto reserved only for outright revolution. Each and every threat calls for the same massive response, whether it be wholesale destruction through atomic attack or a single raid on a nuclear installation or convoy. The report that even a few kilograms of nuclear material is missing would trigger a state of extreme emergency. John H. Barton, a professor of law at Stanford University, predicts in his study of "Intensified Nuclear Safeguards and Civil Liberties" (prepared for the Nuclear Regulatory Commission in 1975) that in the event of a threat entire regions of the country would have to be occupied by what he calls "response forces." Tanks would patrol the streets and helicopters hover noisily overhead. Troops would undertake house-to-house searches without warrant in entire sections of the city. Barton hopes that violations of civil rights would occur only in extreme cases (e.g., the evacuation of densely settled areas after a nuclear plant had been sabotaged). But he fears the worst involving innocent third parties when security forces move in on the terrorists, primarily through the use of "deadly force"—a bureaucratic euphemism for "official murder."

Barton outlines one of the critical situations involving these response forces:

> Finally, dissidents might be seized and detained after a plutonium theft. Detention might be justified as a way to isolate and immobilize persons capable of fashioning the material into an explosive device. . . . Detention could also be used as a step in a very troubling interrogation scheme—perhaps employing lie detectors or even torture. The normal deterrent to such practices—inadmissibility of evidence in court—would be ineffective under the conditions of a nuclear emergency.

At the same conference of leading jurists and nuclear

experts at Stanford University in October 1975 great skepticism was expressed as to whether certain government agencies, such as the FBI, should be assigned to protect civilian nuclear operations. Such an assignment could be construed as an "unlimited mandate to protect American citizens." Based on prior experience, many experts feared that the FBI might "inflate the level of risk" as an excuse to overreact and thus protect its own interests. The conference's final report noted that "in light of these issues, participants agreed that the conference must deal with the possibility of all civil liberties being swept aside during an emergency."

The most chilling incident of all would be a nuclear explosion without prior demands or warning. This might be the destruction of a government complex—or blowing up the Capitol, as envisioned in Taylor's nightmare. Such an assault might not be as senseless as it would first appear, serving the purpose of "decapitating" the existing social structure with one massive blow. Widespread panic would ensue. Such an attack by unknown terrorists, according to California psychologist Douglas DeNike, could serve as the prelude to a viciously treacherous nuclear assault launched from abroad. No one would know for certain who had launched the attack, making it impossible to mount a retaliatory strike.

With such catastrophes, the already blurred distinctions between internal and external warfare would disappear. David Krieger has sketched out a few scenarios as they might affect the United States:

• A factory in France owned by an American firm is poisoned with plutonium oxides. Terrorists threaten further attacks against other U.S. firms unless Washington meets their demands.

• Japanese kamikaze aircraft crash into an American breeder plant, melting its reactor core.

• A German terrorist group threatens an American government institution in Europe with nuclear annihilation unless the Dutch government frees certain political prisoners.

• An international conspiracy contaminates American bases in Latin America and Asia with plutonium oxides and threatens to continue its assaults unless the United States withdraws its nuclear arsenal from Europe.

6

In future, employees at nuclear facilities will no longer be checked by human beings, since the latter may be corrupt, careless, or in league with saboteurs. Instead, they will be confronted by three machines in a small neon-lit room. First they will tap out an identification number, wait for a recorded voice to pronounce four ordinary words, and then repeat them. The employee's speech pattern will immediately be compared with his or her "voice print" in the unit's memory bank. If this test is passed, a mechanical "thank you" will move the employee on to the next checkpoint. There his or her signature will be verified in terms of hand movement and pressure applied. Finally a fingerprint machine will scan the employee's hand. Only then will a bright yellow sign signal "IDENTITY VERIFIED"—permitting entry into the plant.

This automatic gatekeeper has proved 98 percent accurate in tests conducted since 1976 at the Pease Air Force Base near Portsmouth, New Hampshire. It is destined to become standard equipment at U.S. nuclear facilities. Only foreigners and women pose problems, the former because their intonation of English is subject to unintentional variations and the latter because their signature is not always the same.

"We must rely on non-human colleagues—that is, on machines—as much as possible at nuclear facilities," explains G. Robert Keepin, nuclear safeguard program director at the famed Los Alamos Labs. "They are not only more reliable, but cheaper."

One can sense the almost childlike enthusiasm Keepin feels for his electronic control mechanism, especially his unique DYMAC (*Dy*namic *Ma*terials *C*ontrol) System that employs dozens of different electronic controls to monitor activity within the facility. For example, these units can trace fractions of grams of uranium or plutonium during every phase of the production process. The data fed into a computer bank can thus keep tabs on even the minutest speck of plutonium at all times.

The DYMAC System has cost millions to develop and shortly will be installed in all nuclear facilities in the United States and, eventually, throughout the world. It is necessary, Keepin claims, because guards, weapons, and special entry checks can only guard against assaults from the outside. "But we must also protect ourselves against pilferage from within by employees intent on setting aside small amounts of material."

Uranium and plutonium also disappear during the manufacturing process itself, collecting like plaque in the pipes and boilers of the plant. It is a residue that could build up substantially over the course of years. The Nuclear Regulatory Commission estimates that from 1968 to 1976 close to 475 kilograms of uranium and plutonium vanished from the facilities under its control. No one knows for certain whether it remained behind as residue or was pilfered. Just tracking down the answer to this riddle has become a science in its own right, a serious business jokingly termed *muffology* (the search for MUF—"material unaccounted for").

A loss of one percent is viewed as inconsequential. With the burgeoning of nuclear facilities such an unverified and uncheckable amount could accumulate to many pounds of plutonium annually. Who knows what has become of it? Perhaps it is being traded on the plutonium black market, sold to some nation lacking nuclear weapons or, even more ominous, to a band of terrorists.

The Office of Technology Assessment apparently recognizes that the black market in special nuclear materials (SNM) is not just a sensational hypothesis, but a serious danger. In its report on proliferation OTA noted: "A market of several hundred pounds of fissile material worth millions of dollars per year seems credible. Although small in comparison to the drug market, it is large enough to interest criminal groups and to have a major impact on proliferation." OTA figures that the major suppliers are employees of reprocessing plants: "If each smuggled out just one gram of plutonium per day (an amount probably too small . . . to detect) he could realize $5,000 per year and maybe several times that."

It is possible that FBI agent Larry Olson's theories about the cause of Karen Silkwood's death are more relevant than was originally assumed. In Olson's opinion, Karen may have stumbled on a smuggling ring within the Kerr-McGee plutonium plant without realizing it while searching for evidence that safety rules had been violated. Jacques Srouji, a journalist based in Tennessee, met Olson there. Srouji stated publicly: "Karen Silkwood must have had information in her possession which not only pinpointed the exact amount of nuclear material missing, but the persons involved as well. She did not know the time bomb she was carrying."

It is also suspected that U.S. government intelligence operations may have been involved in smuggling operations. Howard Kohn reports: "One theory to which some nuclear experts subscribe is that the CIA had diverted this shipment [of plutonium] to Israel." That would explain why the authorities and the courts ordered to investigate the case of Karen Silkwood clearly shirked their duty and gave the appearance of covering up their tracks. For a time, their most worrisome opponent was Congressman John Dingell, who not only held hearings on the Silkwood case with his Small Business Committee, but also attempted to conduct his own investigation of Karen Silkwood's death with Peter Stockton, a former Budget Bureau economist and weapons analyst.

Dingell directed a strongly worded letter to Attorney General Edward Levi, demanding an immediate and uncensored explanation of the FBI's and the Justice Department's peculiar behavior. By doing so, he may have pushed matters too far. With the deadline approaching that Dingell had given for a reply to his letter, a call girl in Detroit suddenly named Dingell as one of her clients. The resultant scandal stopped Dingell's investigation. Not quite six months later the Democratic caucus in the House took the chairmanship of the committee away from him.

Granted, much of this sounds like a contrived whodunit tragically out of kilter in a serious discussion of the hazards of nuclear proliferation. But the fact remains that the dark side of the "nuclear enterprise" very much resembles an appalling leaf out of a contrived mystery plot. Serious observers and analysts are inclined to dismiss it all as beneath their dignity. In so doing they only serve the cause of those who would see government action slip into a morass of deception and not being accountable to the law.

John Conran's accusations against NRC safety measures mentioned previously raised the distinct possibility, during an investigation of the charges by the Commission, that certain government agencies may have been involved in the illegal spread of fissile material. It is obvious that the disappearance of 400 pounds of uranium from the NUMEC facility in Pennsylvania in 1965 was known to both the intelligence community and NRC. But the facts in the case were deemed so "ultrasensitive" that only a few top administrators were ever given a briefing.

Insiders think that something of a "gray market" was at work here—with the knowledge and the assistance of the U.S. government—in favor of a foreign power. For reasons relating to international politics the transfer of material could not be done openly. The circumstances may have been similar to the disappearance of the *Scheerberg*, a ship carrying a load of uranium. When this vessel finally docked, all the uranium on board had vanished.

Paul Levinson reported the *Scheerberg* case at the Conference for a Non-Nuclear Future in May 1977, seeking to shed some light on the grave implications of nuclear politics in international affairs. Given his work with Senator Abraham Ribicoff and his investigations for the Committee on Government Operations, Levinson had a good overview of the problem.

Levinson's voice was tinged with pessimism when he spoke to me. "If we ever find out sometime where the damned stuff has gone, it might be too late. It's really monstrous how we are building up an industry that produces tons of highly dangerous, unstable material, only a few grams of which could work untold havoc. And even if it never came to that—think of the torrents, the avalanches of anxiety this damned new technology has unleashed on the world!"

Friedrich Hacker, a pioneer in the analysis of terrorist psychology, believes we can wake up from the nightmare of potential nuclear terrorism only if we give up the idea of creating a perfect police state. Second, we should attempt to defuse the basis for terrorism by working toward greater social justice.

I would add a third point: Under no circumstances must the world stock of fissionable material be increased. Every additional kilogram of plutonium increases the appalling danger of its misuse.

SEVEN
Citizens Under Guard

1

"The Nuclear Age. All of us are sitting in this train. It is moving. Who's the engineer? Who planned the trip? Was it the experts? The technocrats? Expertocracy—a weird concept, and yet so much a part of our present-day reality."

These searching questions were posed to thousands of listeners by the Swiss author Otto F. Walter in 1977 at the concluding demonstration against the Swiss nuclear power plant at Gösgen. The protesters had gathered in the shadow of the gray, 105 meter (350 feet) high concrete cooling tower of the plant to demand a "moratorium" in the further development of nuclear power stations.

One young woman, twenty-two-year-old Anna R., remained behind after the other demonstrators had left to reflect on what she had heard. A passing police patrol considered this "bizarre behavior" and forced her into a police wagon without first establishing her identity—the first violation of her rights. At a nearby police station she was stripped and subjected to a body search—the second violation of her rights. In protest Anna R. refused to dress herself again and was thereupon, although no charges were placed against her, held overnight—the third violation of her rights.

The next morning, the "insane woman"—so-called because the Swiss police did not quite know what to make of someone who offered passive resistance to their rough treatment—was taken against her will to a psychiatric clinic. After her identity was established, Anna R. was sent back to her home canton of Geneva, where she was placed under guard in the *Clinique Psychiatrique Universitaire de Bel-Air*. The doctor on duty signed Anna R. into a cell—the fourth violation of her rights. (According to Swiss law, a doctor outside the clinic is required to examine psychiatric patients and sign the papers for their admission.)

Caught in the machinery of repression Anna R. tried desperately to offer passive resistance. She went on a hunger strike, rejected the medicine prescribed for her, and refused to answer the questions of a psychiatrist.

Two days later, the same psychiatrist determined to break down Anna R.'s resistance by subjecting her to electric shock treatment without her own consent or that of a next of kin—the fifth violation of her rights. Under anesthesia Anna R. had a strong electric current sent through her body by electrodes attached to her head, causing convulsions and the interruption of her breathing for several seconds. (It is a fact that this highly controversial form of therapy has caused loss of memory and changes in behavior.)

In the meantime word of Anna R.'s incarceration reached her family and for the first time, a week after her arrest, she was allowed to see someone other than the police or hospital staff. However, the attending psychiatrist

promptly forbade all further visits because they were too up-
setting for the patient. The electric shock treatments continued even as Anna R.'s friends were assured that she was
much better. Ironically, this same psychiatrist also had Anna
R. appear before an examining magistrate, the first step under Swiss law for building a case against her.

But now a protest was raised even within the hospital.
Dr. Bierens de Hahn strenuously objected to the use of electric shock treatment in Anna R.'s case, considering it a
brutal and therapeutically ineffective procedure. For some
time de Hahn had been attempting to persuade the clinic's
chief doctor, Professor René Tissot, to replace his authoritarian methods with the modern, participatory concept of a
"psychiatric community." A colleague joined de Hahn in his
protest. Both were called to the director's office on June 23,
1977, and told to clear out their desks within a week. De
Hahn was told the reason he was dismissed was because "he
had openly opposed electric shock treatments and shown a
lack of solidarity with colleagues who were employing it on
Anna R."

It turned out that the director had been waiting for just
such a chance to rid himself of his two opponents. When Dr.
Bierens de Hahn and his colleague appealed their firing to
the government, it reacted as it had toward the nuclear demonstrators. The authoritarian state sided with the authoritarian director. Later the plucky Dr. de Hahn wrote me that
he had tried to hold his ground "in order to liberate psychiatry from practices that are a scandal and pose a threat to
its integrity. The affair ended up as neither medical nor even
scientific; given the reaction of the Geneva government, it
was simply political."

What of Anna R.? Because of the trauma inflicted on her she had to remain in treatment for some time. Thus, arguing after the fact, the authorities could claim that they had been "right" all along.

2

The link between authoritarian nuclear technology and psychiatry, as evidenced by the case of Anna R., is neither isolated nor unique. Governments will use psychological means to break resistance to nuclear power. In the cause of "reason" they employ a variety of methods, ranging from misleading propaganda to gain "acceptance" to overt brainwashing.

According to the testimony of the Swiss chemist, Dr. Bruno Ferrini, police used not only CN (chloroacetonphenon) but possibly another gas that induces personality changes on demonstrators at the Gösgen nuclear power plant on July 3, 1977. Tests on residues found in the vicinity on cherries and barley seemed to confirm this suspicion. Of course, the police routinely denies such reports. But it is a fact that the Basel police department warns in one of its training manuals: "The general view is that the above-mentioned gases are effective, but harmless. This is not the case. If we are forced to use them we must recognize that those affected might be fatally poisoned."

CN, a gas the police admit they used on the Gösgen demonstrators, was first employed by the Americans on the western front in 1918. The League of Nations condemned its use in 1925, but it has since been used against protesters in both France and Northern Ireland. New, non-lethal weapons for use in civil disturbances are continually being tested, just as are weapons systems for use in war. For example, the British experimented with an ultrasonic device for the first time recently in clearing an occupied university building in Birmingham. The barely audible sound waves proved "quite satisfactory," causing a loss of balance among those at whom it was directed.

In Aberdeen, Maryland, the U.S. Army's Human Engineering Laboratory specializes in the development of non-lethal weapons. Among the 34 weapons in its arsenal to tame recalcitrant citizens are:

• A dart gun that immediately paralyzes its victim with a drug injection.

• The "instant banana"—a fluid so slippery that it makes it impossible to walk or run on areas where it has been sprayed.

• TAESER, a gun that fires two barbs attached to thin wires into a victim's clothing or skin. A powerful electric current triggers immediate unconsciousness.

Common to all these weapons is the fact that they are designed not for war—but for use against a nation's own citizens. They immobilize rather than kill. Their sole purpose is to stop demonstrations before they escalate into bloodshed and possibly civil war.

The Council for Science and Society in London discussed the alarming growth of such weaponry in a report (summer 1978) entitled "Harmless Weapons." The council's long-time vice-president and tireless defender of civil liberties, Paul Sieghart, wondered whether increased reliance on such an arsenal by the police might not alter their image as public protectors to one of a hostile force. "And if that happened,"

Sieghart observed, "might not public order become progressively more difficult to preserve?"

Jonathan Rosenhead, a Marxist political scientist at the London School of Economics, recently collaborated in a study of this new technology for political control. The unrest in Northern Ireland had already triggered such developments as the rubber bullet; but, even so, Rosenhead was amazed at how rapid the developments had been since the end of the sixties. In his small office not far from the "mother of parliaments" Rosenhead told me he was convinced that "politicians and economists will attempt to overcome the ever more obvious crisis of a power structure geared to high production and profits with technological fixes—and, if need be, the use of force. The latter must not be too obvious, of course. American political scientists working in the same field have aptly called it 'the politics of an iron fist in a velvet glove.'

"What the power structure wants most of all," Rosenhead continued, "is to quietly reach a point where no one dares disturb its policies. It seeks to make citizens' lives dependent on consumer goods and jobs, coupling this dependency with intensive surveillance. Then, should conflict erupt, the hope is to contain it with limited measures that will deter its spread."

The massive introduction of nuclear power, first without the knowledge of and increasingly against the will of many people, coupled with the rise in terrorism, has created something of an international dress rehearsal for using all these "internal armaments." British social scientists were among the first to point out that the controversy over nuclear power involved not only the environment, but civil rights as well.

Two such scientists, Michael Flood and Robin Grove-White, told me in their London office on Poland Street that they were deeply disturbed by the neglected social and political problems brought about by commitment to nuclear energy. They wonder whether the coming "plutonium economy" will inevitably bring with it the nightmare of a total state as envisoned by George Orwell in *1984*.

Pamphlets both men wrote on the subject in 1976 struck a responsive chord in parliament and the media. I was surprised to learn that Great Britain, long considered an international repository of human rights, had already acquired some of the trappings of an atomic state. Unlike many nations that remain silent as to whether employees at their atomic installations are given security checks, Great Britain publicly acknowledges that all employees at its Atomic Energy Authority, including workers and mechanics, are carefully screened.

Every major atomic facility in Great Britain is subjected to the Official Secrets Act, which means that all employees are forbidden to talk about their work. In 1976 the British were the first to establish a special police force to guard atomic plants, and especially plutonium storage facilities. These special constables have been granted unheard-of special privileges involving the wearing and use of sidearms. Yet all public discussion related to such privileges, as well as to other special rules made for the Atomic Energy Authority, is forbidden. The system of so-called "D notices" that effectively censors the press on military matters has now been extended to anything having to do with atomic energy.

As long ago as the fifties the London *Daily Express* was forbidden to publish the reports of Chapman Pincher, its top correspondent, on construction problems at the Windscale reprocessing plant. Flood and Grove-White maintain that eventually these stiff regulations will be cast as an ever wider net, drawing in numerous individuals having no immediate connection with nuclear facilities. As more atomic plants are built, more and more people will fall within this net of restrictions.

"Anyone who has ever been critical of nuclear policies can count on some attempts at surveillance," one of the pair confided to me. "The security services could check the backgrounds of suspected individuals at even the first hint of civil disobedience. In the near future the number of special constables involved in nuclear security will total at least 5,000 in Great Britain, backed up by an army of investigators outfitted with the latest technology in 'data finding'—a nice euphemism for spying on one's fellow citizens."

3

One of the aspects of the technocratic state, regardless of political persuasion, is its absolute mania for collecting data on individuals. This is true for both government and private industry. But even those most intimately involved know only those few meshes in their own particular corner of the net. How large is the net? Is there an "information elite" that can draw this mass of data into one cohesive whole? Probably not, since interdepartmental jealousy separates the investigators like so many watertight bulkheads in a ship. Perhaps that is the reason for frequently contradictory decisions in even the most centrally structured states. There is a vacuum at the heart of the bureaucracy. It serves a useful purpose by allowing responsible individuals to plead ignorance in embarrassing situations. For the mighty, at least as important as the "right to know" in security matters is the right *not* to know. It provides a convenient excuse for ignoring sinister operations outside one's immediate narrow field of competence.

Spurred on by fears of nuclear terrorism, industrial states are likely to consider coordinating their data banks on individuals into a comprehensive system of enormous proportions. Nuclear power thus triggers the expansion and concentration of governmental surveillance. The implications are readily discernable in discussions presently going on in Anglo-Saxon nations, with their liberal tradition forcing a public look at what all of this means to individual rights. In the United States a 661-page report by the National Advisory Committee on Criminal Justice, published in March 1977, suggested that emergency legislation be prepared in view of the high vulnerability of our technological civilization. In an emergency these laws could suspend congressional debate or even review by the Supreme Court. Further, the report urges that officials be shielded from civil or criminal court proceedings if their actions under these emergency laws violate the rights of others.

At a conference on "The Impact of Intensified Nuclear Safeguards on Civil Liberties" sponsored by the U.S. Nuclear Regulatory Commission at Stanford University in October 1975 (mentioned in Chapter 6) the entire range of possible violations was discussed by lawyers and technical experts. One observation, in particular, illuminated like a bolt of lightning the ominous role nuclear power could play in converting a democratic nation into a totalitarian atomic state. All at the conference agreed that the "Pu factor" (plutonium) was the "first legitimate justification" for enacting standby legislation requiring a comprehensive system of government surveillance. Thus the dangers of nuclear development will serve as justification for a process that has already been set in motion. What better excuse to violate human rights than this boogeyman set up to serve the purposes of a technocratic state?

In time, laws designed to deal with a crisis—such as an outbreak of terrorism—will become the norm for what is in fact a permanent state of crisis. The watchword will be: Protect at all costs the source of energy. Atomic power continually under threat leads to a permanent state of siege. It brings about harsh new laws to "protect the people." It encourages denunciation of atomic energy opponents and environmentalists as a "precautionary measure." In the end, it will justify everything—from the mobilization of thousands of policemen against demonstrators to the body search of all those arrested simply for exercising their rights as citizens.

All of this begs one question: Is it not the very power it gives to central government what makes atomic power so attractive to the "establishment"? Certainly the economic benefits of this "new energy source" do not begin to justify such a massive transformation of our society.

4

The enormous investment in internal security is in itself an economic benefit of sorts, of course. Has it not developed into a field almost as lucrative as the traditional merchandising of arms? Just as nations want the newest, most expensive weapons, so today's police want the *newest and best* equipment to "maintain internal security." One authority estimates that 2 billion marks were spent in 1975 in "free Europe" on this newest of growth markets: security. The future seems just as secure; expenditures by 1980 are estimated by the West German newsletter "Sicherheitsreport" at 3 billion marks. By 1985 this should rise by another billion marks, with the best long-term prospects being in West Germany. Frost & Sullivan, a market research company, predicts a rosy future for the field of security technology.

Here are some of the newest items available in 1976: laser beams to scan identification cards; exterior lighting with independent circuitry; long-range infrared heat detectors; automatic controls of visitors; a new Sphinx 2 locking system; King Pin Lock theft protection; pictogram plans for rapid evacuation of buildings; bullet-proof polycarbonate plates; a chemical club; security equipment for dispatchers; a magnetic device to protect against sabotage; subterranean nuclear facilities; and a bug-proof listening device.

Wait. There is more: caloric sensors that can detect any warm-blooded body moving past a fence; devices that read voices like fingerprints; scramblers to prevent bugging of phone calls. Also available are a mobile command post and television cameras—the latter originally intended to monitor traffic, but now used by many police departments to observe people.

Just as the Spanish Civil War provided an experiment to test new weapons of war, so Vietnam and Northern Ireland afforded an opportunity to test on living subjects new ways of combatting insurgents, saboteurs, urban guerrillas, and demonstrators. Police and plant guards in many countries have benefited from these experiments. The technology to control human beings took a quantum leap forward.

While entire segments of the economy have been in recession since 1973, it is boom times for the security business, based on continuing innovation and high sales. We thus have a Security Age to supplant the now dormant Space Age, which, after initial high investments, no longer provides much economic stimulation.

5

In the United States the development of "counter-terrorism technology" resides largely in three research centers. They all happen to be well-known laboratories that first made their mark in the development of nuclear weapons: Los Alamos, Livermore, and Sandia. Now their primary emphasis is on designing an arsenal for the "internal front."

The LASL Safeguards Group (Los Alamos) is working on advanced material accounting systems. Their objective is to reduce to a bare minimum the measurement errors of imprecise bookkeeping that previously failed to detect the diversion of plutonium used in fuel fabrication. This system, mentioned earlier as DYMAC, is designed to prevent the potential theft of plutonium for weapons manufacture. In a similar manner, the Lawrence Livermore Laboratory (LLL) is working on an improved systems study of material control and accounting techniques, BETIMAC, for use in reprocessing plants geared to continuous operation.

Nevertheless, an OTA study on nuclear safeguards concedes that even all these new and improved materials accountancy systems cannot do the entire job. "Physical security, containment, and surveillance will still have crucial roles to play in any effective safeguards system."

In the main, this task is being handled by Sandia Laboratories in Albuquerque, New Mexico, under contract from the Safeguards and Security Division of the Department of Energy (DOE/SS). Developments there are continually being updated in a series of handbooks produced by Sandia's Intrusion Detection Technology Division. These manuals are in great demand not only throughout the nuclear industry, but also in government agencies that consider themselves threatened by "internal enemies." The three texts published thus far are:

1. *The Intrusion Detection Systems Handbook*
2. *The Entry Control Systems Handbook*
3. *The Barrier Technology Handbook*

Already in the works are additional handbooks on locks, seals, and central control systems. Sandia's alarm systems grow ever more sophisticated and expensive. An investment of millions is required merely to install a microwave interior ultrasonic detection system to sound the alert should an intruder penetrate the outer barrier of defenses. The problem with the perimeter of an installation is that the dozens of sensors installed there are subject to false alarms triggered by "innocent people, large animals, blowing debris, etc."

The sensors are there to protect, but must themselves be protected in order to function properly. This problem was outlined in typical security jargon by James D. Williams in *Nuclear Materials Management* (Spring 1978). As Williams sees it, the system can only be made less vulnerable with "tamper alarms, anti-capture circuitry, line supervision capability, and full end-to-end self-test capacity. Installation practices such as overlapping the sensor fields to provide mutual protection for each sensor or the addition of special sensors (point sensors) to protect weak points in the sensor system are also essential considerations. . . . .Typically, an intrusion detection system employs perimeter sensors, building penetration sensors, and interior sensors. Additionally, proximity sensors or movement sensors are often installed on critical or sensitive materials. . . . .This combination demonstrates a safeguards concept known as 'protection-in-depth'; i.e., an intruder must successfully circumvent or defeat each of the protective measures in sequence before access to the protected material or facility is achieved."

Yet even in this array of electronic barriers Manfred von Ehrenfried detects a gap that could be exploited on the marketplace. In a *Nuclear Security Safeguards Newsletter* he inquires of harried nuclear plant managers: "Now don't you think you need a small security computer before your wife gets one for the kitchen and the family use?" Of course the manager who wants to keep up with the state of the art must in addition acquire one of these magnificent cathode ray display units, a dazzling multicolored gadget that flashes red for alarms in action, green for alarms secured, yellow for areas in access, and blue for recording guard procedures.

Only one thing is lacking in this Rube Goldberg world of nuclear age security: A gadget that meshes the whole thing with its human components, the guards. In June 1975 two guards at the Three Mile Island nuclear plant carried their plight to the press. Among their complaints: "TV cameras that are supposed to monitor the site fence don't function all the time. . . . Guards frequently turn them off. . . . Records are falsified. . . . Records show that guard sergeants made post checks when, in fact, the sergeants did not make these checks at the times recorded."

The prevalence of the human fallibility problem has given rise to such companies as the W. H. Brownyard Company of New York that offers insurance against "wrongful acts of hired security agencies." The company jars readers of the trade journal *Security Management* awake with a bold headline: "BUYERS OF SECURITY SERVICES BEWARE!"

Can better training eliminate these deficiencies? In the United States new schools have cropped up to do just that with books, films, lectures, and workshops. One of the latter offers such features as:

> Interview and Interrogation
> Emotional Inroads and Breakthrough
> Body Armor, Weapons, and Ammunition
> Optimizing Individual Guard Performance
> Security Awareness
> Dealing with Terrorism
> International Intelligence Operations
> Hands-On Methodology (to assist
> and advise those engaged in the design,
> engineering, building, and securing of
> nuclear power plants)

What about manpower? The newsletter of the American Society for Industrial Security (ASIS) published in Washington, D.C., reported:

At present 30,000 young adults between the ages of 15 to 21 are actively enrolled as law enforcement explorers. . . . The officers and directors of ASIS believe that ASIS members will be particularly interested in activities involving interchange between the field of security . . . and the young adults who are active in Law Enforcement Exploring.

6

A number of these future law enforcers will almost certainly work as informers and undercover agents at some later date. As Russell Ayres warned in his provocative paper that appeared in a 1975 edition of the *Harvard Law Review:* "Plutonium provides the first rational justification for widespread intelligence gathering against the civilian population . . . . In the past federal courts have taken a skeptical view of attempts to justify spying on national security grounds, but with the very real threat of nuclear terrorism in the picture, the justification is going to sound very convincing."

At the previously mentioned Stanford conference the problem was even more clearly spelled out: "It is likely that Pu use would create pressures for infiltration into civic, political, environmental, and professional groups to a far greater extent than previously encountered and with a greater impact on [free] speech and [its] associational rights." The pressures would be exerted not only on the few thousand employees in sensitive nuclear energy jobs, but against millions of citizens who might have expressed criticism at one time or another of political and industrial developments.

Intelligence operations in this nuclear age are viewed as "the first and most important line of defense." In the United States more than thirty federal agencies in the intelligence community are involved in planning how to cope with a nuclear emergency, as was reported in a critical study prepared by the Citizens' Energy Project in Washington, D.C., under the ominous title, "Nuclear Power and Civil Liberties: Can We Have Both?"

Not only federal agencies, which have resumed wire-tapping and spying on their own citizens with renewed vigor after a temporary hiatus due to Watergate, but private investigation firms are also involved in the effort. The woman who runs Research West in San Francisco refused an invitation to appear before Representative John Moss's congressional committee. Apparently she would rather risk a contempt citation than reveal any of the "dirty tricks" used by her firm.

Reports from all over the United States as well as Great Britain, France, Italy, Spain, and (particularly) West Germany tell of the devious and at times criminal actions of informers who, as ostensible "defenders" of law and order, subvert justice. They help to undermine the very basis of human understanding that is built on trust and respect for individual rights. (I refrain from recounting any of these rather lurid stories here because they would only detract from the larger tragedy of the dangerous directions in which nuclear energy and its demands for security are moving us.)

Still, it is my contention that the "internal arming" of America will lead to a "police-industrial complex" to rival the "military-industrial" one. Swollen out of all proportion by the thousands of people involved, it stands to threaten the ideals of 1776 that left the American nation a legacy of representative government and hard-won rights for the common man.

Fascism and Nazism had scarcely been destroyed when a small group of political and economic power brokers in both

East and West took control of governments, seemingly with the acquiescence of those they governed. The legitimate fear of nuclear annihilation appears greater today than does the love of freedom. The perceived need for protection has replaced the desire for participation.

The hard question must be asked whether those who profit the most from this public anxiety are not themselves involved in exacerbating the tensions. Are they perhaps not pumping up the significance of their opponents in order to justify ever more investments in their gigantic safeguard apparatus? Isn't it perhaps possible that they somehow instigate terrorists—the suspicion was often voiced in relation to the Moro kidnapping—and then afford them protection from arrest? European history of the nineteenth century teaches us that such tricks and provocations were often used by the police. In those days, attempts were made to stop the workers' movement by inciting terrorists into attacks against royalty—a cynical maneuver that often paid off with devastating effect.

Where this nuclear safeguards mentality is taking us is revealed in a comment made by Professor Alvin M. Saperstein, a political scientist at Wayne State University in Detroit. In *The Bulletin of the Atomic Scientists* Saperstein urged a "new Inquisition" to protect human health—a necessity forced on us, he wrote, by the known risks of atomic power. It is hard to figure out whether Saperstein was being satirical or meant to be taken seriously.

But then again it may be an idea whose time has come. At the Stanford conference someone asked whether the use of plutonium would not heighten the public's alarm over still more government intrusion. The answer given: Experience has shown that people are prepared to accept "less absolute privacy" in view of the hazards of a technologically complex society.

The last decades have seen a variety of laws passed to protect man from the folly of his technical advances. After a time they were all accepted as a matter of course. It started in small ways, such as limiting the rights of pedestrians to use the road in the face of automobile traffic. This has escalated to the searching of one's person and luggage before boarding an aircraft, an invasion of privacy that was accepted with scarcely a ripple of protest. Similar methods are already being used before we are permitted to enter certain office buildings.

The question is whether these increased restrictions will not push people over the brink of toleration. Security experts are much concerned with the possibility of "incidents" or "explosions" in their new technology. But they underestimate the possibility of social "explosions" that appear almost equally inevitable.

The human factor adds a dimension that is even less predictable for the new tyranny than a reactor run amok.

EIGHT

The Soft Path: A Prospectus

1

It isn't often that members of a profession turn against their colleagues by warning the public about them. That is precisely what happened in August 1977, when 28 prominent physicists from 12 nations stated their position on the politics of atomic energy after a conference at the Scuòla Internazionale Enrico Fermi on Lake Como:

> The greatest liability is that the discussion of these problems is not really public, but dominated by an elite of experts . . . . Proponents of nuclear energy seek out only those scientists who are committed to public nuclear power programs . . . . We call on the public to be very critical of the views of these experts and not to follow blindly the statements of those who claim to know more.

This protest against a zealous "priesthood" which serves a dogmatic science is one of the most significant results of the resistance to nuclear power plants. It is clear evidence that the atomic debate has split even the scientific community with consequences reaching far beyond the Lake Como Conference. At stake is not only future energy production, but how we will be governed. It ranges far beyond a particular technology to include the entire structure of our highly industrialized society. The larger question looms: Is mankind still to be served by a system of technological progress based on domination and exploitation?

The disturbing implications of this question have unleashed an international mass movement like no other. Its supporters cut across national, class, and political lines. There is no central direction, no formal program or fixed structure. It is an ever swelling tide, symbolized less by a monolithic creed than by the many streams of discontent and frustration that feed into it.

As yet there is still doubt as to its meaning—whether it will last or have the ability to topple the well-organized institutions of the present power structure. But no one can deny that this political force outside the normal channels of influence has already had enormous impact wherever it appears. It has upset the timetables of bureaucratic planners and industrial management. No longer can nuclear power plants be built with impunity according to fixed schedules. Cost projections soar. Gone is the early optimism of the atomic age.

The "nuisance value" of this public protest is by no means negative. Rather, it can be likened to a pain experienced by the body which, if properly assessed, will lead to a more sensible life-style. Just what such a life with dignity means is the subject of endless discussions, not only by the young still seeking their place in society, but also by those already established and now searching for a deeper reason for their existence. The resistance movement has brought the latter out of their isolation and shaken up their daily routine. They identify with the protesters to the extent that they, too, are looking for values which will have meaning for their lives and to which they can subscribe.

2

Some see the movement against nuclear energy only as a negative act of protest. In fact, the people involved in it are not just protesters *against;* most of all, they are committing their lives *for* something.

The farmers of Wyhl, Saint Laurent, Kalkar, and Brokdorf were demonstrating for the preservation of their way of life. The workers at La Hague went on strike for the sake of their endangered health. The Clamshell Alliance protesting the nuclear power plant at Seabrook is fighting for the environment. Japanese, Basques, Italians, and Dutch went on hunger strikes for the sake of future generations threatened by nuclear pollution. Australian dock workers struck for the sake of aborigines threatened by uranium mining. The demonstrators at Gösgen, Barsebeck, and Zwentendorf were for maximum public involvement in how tax monies were to be spent on massive power projects.

To make their point, they have taken a leaf from the counter-culture and student movements of the past. They have succeeded in establishing realistic objectives that can count on ever wider support. In my contacts with anti-nuclear groups throughout the world I met many professionals who now share the views and hopes that a few years ago were held only by fringe elements. This is especially true of the mounting skepticism expressed by professional people.

It is wrong to claim that the protests of the sixties are dead. The motivation that triggered them has simply penetrated other social groups and is expressing itself now in different forms. In various demonstrations I met architects, lawyers, doctors, construction workers, ministers, farmers, fishermen, druggists, book sellers, public employees, salesmen, journalists, nurses, teachers, mechanics, actors, and printers. It is significant that they have all come into the open to stand on common ground. No one would say that only "destructive spirits" make up the protesters against atomic plants if they could have heard, as I had, an engineer and an organ builder sharing experiences in their united struggle against the nuclear plant at Esenshamm. Once again those who had been separated by the economic development of the past century can join together in a cause that unites us all.

It is impressive how these people from varied backgrounds will take the trouble to study the implications of nuclear power and master complicated technical data. They weigh them critically and then apply them to their own situations. They are more powerfully motivated to learn than the average citizen and quickly assimilate the new material. Often they come better prepared to discussions than do the local politicians and representatives of industry. And they can no longer be brushed aside with slogans and weasel words.

At such confrontations it is interesting to listen not only to what is said, but how it is said and to watch the faces of the participants. Invariably, supporters of nuclear power plants present an image of casual restraint, boredom, aloofness, convoluted "objectivity," and smug superiority, with scarcely a trace of warmth or empathy. The faces of their opponents are lively and attentive; full of enthusiasm and always quick to laugh.

3

From the many discussions, letters, and experiences I have had with this new international movement I have attempted to distill the essence of its aspirations.

Above all, there is a devotion to a more modest life-style —less is better. This concern came with the realization that our earth's resources are limited and that the extravagance of the industrialized nations cannot go on forever. We face a future of shortages rather than unlimited wealth. Those who make do with less may be laughed at today; but, according to the Stanford Research Institute, by the turn of the century (at the latest) they will serve as role models for all of us.

Closely tied to this ideal is the search for justice. The movement takes seriously the huge gap in the standards of living between the developed and less developed nations. It can tolerate neither the exploitation of the Third World nor the disguised "blessings" of their own economies that seek to export a life-style that already weighs so heavily on the industrialized world.

At the heart of this discontent is the long-standing destructive relationship between modern technology and the environment. Early feelings of helplessness have given way to a recognition that all of us are guilty to some extent of contributing to this condition. The growth of the environmental movement has inevitably led to opposition to nuclear power. There is simply no convincing argument that the latter can ever serve the cause of environmental protection. One of the prime objectives of the movement is to accept responsibility for the preservation of our threatened planet earth—this "blue marble," as one of the astronauts called it from his vantage point in space. It is a fundamental objective based less on fuzzy idealism—as has often been assumed—than on a necessity that was ignored by earlier international movements.

The movement also seeks to make it politically acceptable to consider such values as diversity, beauty, and creativity . . . values that have been ignored in the pursuit of ever higher productivity. It is no accident that its adherents encourage street theater, folk music, the artist and poet that is in each of us, as the basis for a new culture to replace many of the traditional art forms that have grown stale.

There is a force at work here that seeks to give full expression to our humanity. Never have I experienced such feelings of spontaneous good will, friendship, solidarity, and empathy as at the demonstrations of nuclear opponents. There were the small things—the ways in which strangers talked with one another, the laughter, the willingness to be close together, the unabashed tears, whether in sorrow or rage; none of these has ever been conveyed, to my knowledge, by television.

You have to be there to feel it. Within a demonstration are countless groups linked by a common denominator, yet so much more diverse than the huge organizations and faceless bureaucracies we have become accustomed to dealing with. The many real differences between these groups should not be seen as a sign of weakness, but as proof of the movement's vitality.

Participation is the political lifeblood of the movement. People learn from and listen to one another; policies are forged through discussions rather than from prepared texts. Everyone is an indispensable "expert" for his own needs. It is this sharing in the creative process that requires time, a rare commodity in a society geared to speed and dominated by the clock.

At its best, the movement strives to unlock the individual creativity that is in each of us. It has no leaders or opinion makers to force their will on others. It seeks a constant stream of energy from many hearts and minds—the power of human creativity rather than atomic power.

4

A love of beauty and justice; imagination, involvement, and a reasonable approach to our relationship to nature— these are among the *positive* values put forward by the movement against the new tyranny. Is there any hope that such dreams will ever become a reality? Or are the "realists" right —the ones who proclaim that sheer necessity forces us on to the "hard path" of ever more technology and energy? Not to do so, they say, would lead to an international crisis of shortages, the collapse of civilization, and a bloody fight over the remnants.

It is this inability to alter their mindset that marks adherents of the hard path. They find it impossible to admit earlier mistakes, to accept small losses to avoid an even greater calamity in the future. Theirs is the pursuit of an end regardless of the consequences.

What chance for the "soft path"? Can it be mapped out even today, anticipating possible crises and dead ends—a guide to the future that would be persuasive to all? Can it find its way through the cracks in today's power structure?

As remote as such a possibility seems at the moment, there is evidence that new, soft technologies can replace parts of the hard path with energy derived from the sun, wind, tides, photosynthesis, and a variety of other techniques. Previously dismissed by the establishment as uneconomic, these environmentally benign sources of energy are at last being pursued vigorously. It is remarkable how much creativity has been invested in such ventures since the 1973 oil crisis, a development that bodes well for the future. Never before have so many useful inventions been made and patented in the field of energy production in so short a time.

Proponents of nuclear energy insist that these new technologies will not be ready in time to fill anticipated gaps in the late 1980's. But new studies have revealed how much this pushing of the panic button by nuclear energy supporters is off the mark.

Another hopeful sign is that the views of supporters of the soft path are increasingly gaining broader acceptance. The electronic media can play a decisive role in bringing new concepts into public discussion and discrediting policies that are no longer tenable. One example of this increased awareness, thanks to the images portrayed by television, is the concern for the environment. Shared by a broad spectrum of the public, it has become a powerful political factor. A new perception of the world is emerging, carried forward by our youth—one that the existing political parties and labor unions will find difficult to ignore.

Despite these hopeful signs, it is still possible that the growth of the new tyranny will temporarily push the nonviolent new international movement into the catacombs. But the technological tyranny is at once more powerful and more vulnerable than earlier tyrannies. In the end, water conquers the rock.

Glossary

ACCEPTANCE—The efforts of government and industry to persuade the public to acquiesce to innovations. In particular, this policy is aimed at removing resistance to nuclear power plants by use of socio-psychological techniques.

ATOM—The smallest component of chemical elements. It consists of a nucleus, which contains almost the entire mass of the atom, and an envelope of electrons that encircles the nucleus in one or more orbits. The nucleus consists of protons (which have a positive charge) and neutrons (which have no charge) of roughly equal mass. The lightest of all elements, the hydrogen atom, consists of one proton and one negatively charged electron. The mass number of an atom is equal to the weight of all particles in the nucleus, while the atomic number is the total weight of only the protons. In a non-ionized atom, or one that is electrically neutral, the number of protons in the nucleus is equal to the number of electrons in orbit. The number and orbital arrangement of electrons determines the chemical properties of atoms.

ATOMIC ENERGY COMMISSION (AEC)—A U.S. government commission established by the Atomic Energy Act of 1946 for research into both military and peaceful uses of atomic energy. The AEC was also charged with responsibility for the utilization of atomic energy, the security of facilities, and the protection of the environment. This combination of two objectives—utilization and control—in one agency's mission was often criticized. In 1975 the AEC

180

was supplanted by the Nuclear Regulatory Commission (NRC) and the Energy Research and Development Agency (ERDA), the latter assuming responsibility for research and development. The ERDA has since been absorbed by the Department of Energy (DOE).

CERTIFICATION HEARING—A nuclear facility can only be certified for construction with the approval of the people living in the area. However, the hearings are an expensive and time-consuming procedure. Since the facilities both benefit and endanger areas far removed from it, attempts are being made to widen the scope of certification hearings. Nuclear opponents in Switzerland achieved a greater voice in such hearings in October 1977.

CONTAMINATION—Radioactive pollution.

CONTROL RODS—Neutron-absorbing elements, such as cadmium, formed into solid rods and pushed deep into the reactor to absorb excess neutrons in the fission process.

CORE—The interior of the reactor in which nuclear fission takes place.

DEACTIVATION—The removal of spent fuel rods and other radioactive "waste" from the reactor.

ECOLOGY—The science that studies the relationship of various life forms to one another (bacteria, plants, animals) within the context of their environment, such as a forest, known as an *ecosystem*.

EURATOM—The European Community's atomic energy agency.

FAST BREEDER—A reactor that creates more fuel than it consumes. The basic fuel is plutonium, with liquid natrium or sodium serving as a coolant. The temperatures in a fast breeder are higher than in a "swimming pool" reactor and the core is more difficult to regulate than in the latter type reactor. The dangers of an accident are proportionally greater, complicated by the fact that sodium is highly reactive with air and water. The meltdown of a core could accumulate so much plutonium in a compact mass that a nuclear explosion could result. Such an "atomic bomb" explosion would be impossible with a uranium reactor.

FISSION—The splitting of certain heavy atoms, such as uranium-235 and plutonium, by bombarding them with neutrons. The split nucleus forms two halves of a roughly equal mass, but of a total mass less than the whole. The missing mass is expended as energy.

FISSIONABLE MATERIAL—Isotopes that serve as reactor fuels; in the main, U-235 and plutonium.

FISSION PRODUCTS—Nuclear fuel residues; highly radioactive materials that eventually decay into non-radioactive or stable elements.

FUEL RODS—Long, thin metal tubes filled with pellets of fuel, or with pressed and sintered Uranium-235 or plutonium.

FUSION—The fusing or uniting of atoms of light elements, such as hydrogen, helium, or lithium, into heavier elements. In the process part of the mass is released as energy.

GAS CENTRIFUGE PROCEDURE—A method used in Germany and other European countries to enrich uranium.

GAS-GRAPHITE REACTOR—A reactor type widely used in Great Britain and France, using a gas (such as carbon dioxide or helium) as a coolant, and graphite as a moderator. Its advantage is that it can use natural uranium instead of enriched fuel.

GEIGER COUNTER—A device used to measure radioactivity, consisting of a thin metal cylinder the axis of which is crossed by a fine wire. A high electrical potential is created between the cylinder and wire. Radiation creates an ionized field which triggers a short burst of power that registers on a counter.

HALF-LIFE—The measure of the average lifetime of a radioactive substance or isotope. One half-life is the amount of time it takes for any quantity of the substance to lose half its radioactivity, either by decaying into another element or isotope. Half-life may be measured in fractions of seconds or millions of years. After 10 half-life periods only a thousandth (2^{10}) of the radioactive material remains; after 20 half-life periods only a millionth; and so forth.

INTERNATIONAL ATOMIC ENERGY AGENCY (IAEA)—A UN committee for furthering the peaceful uses of atomic energy and for enforcing the Non-Proliferation Treaty. Headquarters are in Vienna.

INTERNATIONAL INSTITUTE OF APPLIED SYSTEMS ANALYSIS (IIASA)—An institute to research important international problems with the aid of mathematical models. It is sponsored by science academies of several western and eastern nations, with headquarters in Laxenburg near Vienna.

ION—An electrically charged atom or molecule.

IONIZED RADIATION—Radiation triggered by the electrons of an atom, thereby giving the latter an electrical charge. It includes radioactive, roentgen, and certain ultraviolet radiation.

JET SEPARATION PROCEDURE—A process for enriching uranium developed in Germany.

MELTDOWN—A major nuclear accident in which the fuel rods melt, due to insufficient cooling in the reactor.

MODERATOR—A substance to brake the velocity of neutrons so that the fissionable material will most readily accept them, split apart, and trigger a chain reaction. The most commonly used moderators are graphite, and light and heavy water.

NEUTRON—An electrically neutral particle that can decay into a positively charged proton and a negatively charged electron.

NEUTRON CORROSION—The damage done to reactor components by the constant bombardment of neutrons. Similar to rust in its effect, the corrosion hastens the decay of metals that are already subjected to radiation, high pressures, and extreme temperatures.

NON-PROLIFERATION TREATY—An international treaty to prevent the spread of nuclear weapons as a result of an ever-widening use of atomic energy for peaceful purposes. Enforcement is in the hands of the International Atomic Energy Agency (IAEA). Some nations already in possession of nuclear weapons, such as France and China, or with the capability of producing them, such as India, Pakistan, South Africa, Brazil, Argentina, Israel and Egypt, have not subscribed to or ratified the Non-Proliferation Treaty. Signatories among the original nuclear powers, the United States, the Soviet Union, and Great Britain, are exempt from the enforcement procedures insofar as they do not voluntarily open their facilities to inspection. The export of nuclear components by signatory nations is permitted only if specified prohibitions against military utilization are met.

NUCLEAR FUEL CYCLE—The fuel used in atomic energy follows a cycle of mining and enrichment of the uranium, its conversion into pellets, active use in fuel rods within the reactor (usually 6–8 months), removal of expended fuel, and the reintroduction of the fuel into the reactor. Extraction of the waste product, plutonium, provides a nuclear fuel that can be used in reactors or nuclear weapons.

NUCLEAR PARKS—The concentration of reactors, reprocessing plants, and enrichment and deactivation facilities in a small area, preferably in remote regions or islands in the ocean. This arrangement would minimize the dangers to populated areas both in the event of a nuclear accident and during the transportation of nuclear materials. The marketable product of such parks could be hydrogen, which would be delivered as fuel to consumers.

NUCLEAR REGULATORY COMMISSION (NRC)—The U.S. agency charged with policing nuclear facilities after the 1975 dissolution of the Atomic Energy Commission (AEC).

PLUTONIUM (Pu)—A chemical element not found naturally on earth. It is created in reactors fueled with U-238 and is itself a dangerously radioactive fuel for use in reactors and as the active ingredient of atomic bombs.

PLUTONIUM ECONOMY—An economy based on fast breeder reactors in which plutonium is the principal fuel.

PROLIFERATION—The spread of nuclear weapons to nations that do not possess them now.

RADIATION-PROOF GEAR—Protective devices that shield the body from Alpha and Beta radiation, but afford little protection against Gamma or Neutron radiation.

RADIOACTIVITY—The spontaneous decay of an unstable atomic nucleus. It is the energy released as the nuclei convert to another isotope or element. Alpha rays are helium nuclei (2 protons and 2 neutrons tightly bound); Beta rays are electrons released when a neutron decays into a proton and an electron; and Gamma radiation is made up of X-rays and similar emissions of the electromagnetic spectrum. The product resulting from such radiation is often itself unstable, which leads to still further radiation. Unlike fission, the rate of radioactive decay can be measured precisely on the half-life scale.

REPROCESSING—The reworking of nuclear "wastes" from a reactor. The material that has absorbed too many neutrons is separated for permanent disposal, while the uranium is repackaged as fuel, as is the plutonium, which can also be used in the manufacture of atomic weapons.

RISK RESEARCH—A method developed by Charles Starr early in the seventies to set up probability scales for the occurrence of accidents in technical systems. In a joint project with IIASA and IAEA, H. J. Otway broadened the concept of risk research through the use of psychological parameters.

SAFEGUARDS—The Non-Proliferation Treaty mandated the international control of nulear fuels and technology. Steps taken in this endeavor are commonly referred to as *safeguards.*

SNM (Special Nuclear Materials)—Weapons-grade materials produced through human intervention, such as enriched uranium or plutonium.

"SWIMMING POOL" REACTOR—One of the most commonly used reactors in the United States, West Germany, and other countries, in which water serves as both a moderator and a coolant for a unit fueled with enriched uranium. The water may be *light* (H_2O) or *heavy* (H_2O_2), in which a heavy hydrogen atom of one proton and one neutron (deuterium) is linked to an oxide.

TECHNOCRACY—The subservience of democratic institutions and ideals to the demands of technological requirements.

TRANSURANIUM ELEMENTS—New elements not found in nature that are created in nuclear reactors or accelerators. Their nuclei contain more protons than uranium and include such elements as plutonium, neptunium, and curium.

TRITIUM—A type of water created in light-water "swimming pool" reactors with a hydrogen nucleus consisting of one proton and two neutrons.

URANIUM ENRICHMENT—A process that increases the quantity of U-235 in natural uranium (which is comprised of only 0.7% U-235) to the 3% level required for "swimming pool" reactors. The original American method involved bubbling the uranium compound through porous tubes, whereby the gaseous U-235 seeped through the latter. The procedure must be repeated some 3,000 times until the required 3% level is reached. The jet separation and gas centrifuge procedures achieve the same result by spinning U-238 atoms into a matrix of U-235. The separation method requires 100 evolutions, while the centrifuge procedure requires 10–30 steps to achieve the necessary enrichment. Still further enrichment yields uranium of a potency sufficient for nuclear weapons.

Books for
Further Reading

Alexander, Peter. *Atomic Radiation and Life*. Rev. ed. Baltimore: Penguin, 1965.

Berger, John J. *Nuclear Power: The Unviable Option*. Palo Alto, Calif.: Ramparts Press, 1976.

Bookchin, Murray. *Our Synthetic Environment*. New York: Harper & Row, 1974.

Breach, Ian. *Windscale Fallout: A Primer for the Age of Nuclear Controversy*. Harmondsworth, Middlesex, England: Penguin, 1978.

Brodine, Virginia. *Radioactive Contamination*. New York: Harcourt Brace Jovanovich, 1975.

Bryerton, Gene. *Nuclear Dilemma*. New York: Friends of the Earth/Ballantine, 1970.

Caldicott, Helen. *Nuclear Madness*. Brookline, Mass.: Autumn Press, 1978.

Carson, Rachel. *Silent Spring*. Boston: Houghton Mifflin, 1962.

Cervenka, Zdenek, and Barbara Rogers. *The Nuclear Axis: Secret Cooperation Between West Germany and South Africa*. London: Julian Friedman Books Ltd., 1978.

Cochran, Thomas B. *The Liquid Metal Fast Breeder Reactor*. Baltimore: Johns Hopkins Press/Resources for the Future, 1974.

Commoner, Barry. *Science and Survival*. New York: Viking, 1966.

———. *The Closing Circle*. New York: Knopf, 1971.

———. *The Poverty of Power*. New York: Knopf, 1976.

Curtis, Richard, and Elizabeth Hogan. *Perils of the Peaceful Atom*. New York: Ballantine, 1969.

Ebbin, Stephen, and Raphael Kasper. *Citizen Groups and the Nuclear Power Controversy.* Cambridge, Mass.: MIT Press, 1974.

Energy Policy Project of the Ford Foundation. *A Time to Choose: America's Energy Future.* Cambridge, Mass.: Ballinger, 1974.

Faulkner, Peter, ed. *The Silent Bomb: A Guide to the Nuclear Energy Controversy.* New York: Vintage/Friends of the Earth International, 1977.

Ford, Daniel F., and Henry W. Kendall. *An Assessment of the Emergency Core Cooling Systems Rulemaking Hearings.* Cambridge, Mass.: Union of Concerned Scientists; and San Francisco: Friends of the Earth, 1974.

Freeman, S. David. *Energy: The New Era.* New York: Vintage, 1974.

Friends of the Earth. *Stockholm Conference ECO.* Vols. 1–3. San Francisco: Friends of the Earth, 1972–1975.

Fuller, John G. *We Almost Lost Detroit.* New York: Reader's Digest Press/Crowell, 1975.

Glasstone, Samuel. *Source Book on Atomic Energy.* 3rd ed. New York: Van Nostrand Reinhold, 1968.

Gofman, John W., and Arthur R. Tamplin. *Poisoned Power: The Case Against Nuclear Power Plants.* New York: New American Library, 1974.

Gyorgy, Anna, and friends. *No Nukes: Everyone's Guide to Nuclear Power.* Boston: South End Press, 1978.

Hays, Denis. *Rays of Hope: A Global Energy Strategy.* New York: Norton, 1977.

Holdren, John, and Philip Herrera. *Energy: A Crisis in Power.* San Francisco: Sierra Club, 1971.

Inglis, David. *Nuclear Energy: Its Physics and Social Challenge.* Reading, Mass.: Addison-Wesley, 1973.

Jungk, Robert. *Brighter Than a Thousand Suns: A Personal History of the Atomic Scientists.* Harmondsworth, Middlesex, England: Penguin, 1970.

Kendall, Henry W. *Nuclear Power Risks: A Review of the Report of the American Physical Society's Study Group on Light Water Reactor Safety.* Cambridge, Mass.: Union of Concerned Scientists, 1974.

Knelman, Fred. *Nuclear Energy: The Unforgiving Technology.* Edmonton, Canada: Hurtig, 1976.

Komanoff, Charles. *Power Plant Performance.* New York: Council on Economic Priorities, 1976.

———— et al. *The Price of Power: Electric Utilities and the Environment.* Cambridge, Mass.: MIT Press, 1974.

Laitner, Skip, and others. *A Citizen's Guide to Nuclear Energy.* Washington, D.C.: Center for the Study of Responsive Law, 1975.

188 • THE NEW TYRANNY

Lash, Terry R., John E. Bryson, and Richard Cotton. *Citizen's Guide: The National Debate on Handling Radioactive Wastes from Nuclear Plants.* Palo Alto, Calif.: NRDC, 1975.

Lewis, Richard S. *The Nuclear-Power Rebellion: Citizens vs. the Atomic Industrial Establishment.* New York: Viking, 1972.

Lifton, Robert Jay. *Death in Life.* New York: Random House, 1967.

Lovins, Amory B. *World Energy Strategies: Facts, Issues, and Options.* San Francisco: Friends of the Earth; and Cambridge, Mass.: Ballinger, 1975.

—— and others. *Soft Energy Paths: Toward a Durable Peace.* San Francisco: Friends of the Earth; and Cambridge, Mass.: Ballinger, 1977.

Lovins, Amory B., and John H. Price. *Non-Nuclear Futures: The Case for an Ethical Energy Strategy.* San Francisco: Friends of the Earth; and Cambridge, Mass.: Ballinger, 1975.

Lowrance, William W. *Of Acceptable Risk: Science and the Determination of Safety.* Los Altos, Calif.: Kaufmann, 1976.

McPhee, John. *The Binding Curve of Energy.* New York: Ballantine, 1975.

Metzger, H. Peter. *The Atomic Establishment.* New York: Simon and Schuster, 1972.

Morgan, Richard, and Sandra Jerabek. *How to Challenge Your Local Electric Utility: A Citizen's Guide to the Power Industry.* Washington, D.C.: Environmental Action Foundation, 1974.

Munson, Richard, ed. *Countdown to a Nuclear Moratorium.* Washington, D.C.: Environmental Action Foundation, 1976.

Murphy, Arthur W., ed. *The Nuclear Power Controversy.* Englewood Cliffs, N.J.: Prentice-Hall, 1976.

Nader, Ralph, and John Abbotts. *The Menace of Atomic Energy.* New York: Norton, 1977.

Novick, Sheldon. *The Careless Atom.* New York: Dell, 1969.

Nuclear Energy Policy Study Group. *Nuclear Power: Issues and Choices.* Cambridge, Mass.: Ballinger, 1977.

Olson, McKinley C. *Unacceptable Risk: The Nuclear Power Controversy.* New York: Bantam, 1976.

Patterson, Walter C. *Nuclear Power.* Harmondsworth, Middlesex, England: Penguin, 1976.

Pollard, Robert, ed. *The Nugget File.* Cambridge, Mass.: Union of Concerned Scientists, 1979.

Primack, Joel, and Frank Van Hippel. *Advice and Dissent: Scientists in the Political Arena.* New York: New American Library, 1974.

Reynolds, William C., ed. *The California Nuclear Initiative: Analysis and Discussion of the Issues.* Palo Alto, Calif.: Stanford University, Institute of Energy Studies, 1976.

Sternglass, Ernest. *Low Level Radiation.* New York: Ballantine, 1973.

Thompson, Theos, and J. G. Beckerley. *The Technology of Nuclear Reactor Safety.* Cambridge, Mass.: MIT Press, 1973.

Union of Concerned Scientists. *The Nuclear Power Issue: A Source Book of Basic Information.* Cambridge, Mass.: UDC, 1974.

————. *The Nuclear Fuel Cycle: A Survey of the Public Health, Environmental, and National Security Effects of Nuclear Power.* Rev. ed. Cambridge, Mass.: MIT Press, 1975.

Warnock, Donna, and Ken Bossong. *Nuclear Power and Civil Liberties: Can We Have Both?* Washington, D.C.: Citizen's Energy Project, 1978.

Webb, Richard E. *The Accident Hazards of Nuclear Power Plants.* Amherst, Mass.: University of Massachusetts Press, 1976.

Weil, George L. *Nuclear Energy, Promises, Promises.* Available from the author at: 1730 M Street NW, Washington, D.C., 20036 ($2.00).

Willrich, Mason, and Theodore B. Taylor. *Nuclear Theft: Risks and Safeguards.* Cambridge, Mass.: Ballinger, 1974.

Periodicals

The Bulletin of the Atomic Scientists
Educational Foundation for
   Nuclear Science, Inc.
1020–24 East 58th Street
Chicago, Ill. 60637

CBE Environmental Review
Citizens for a Better Environment
59 East Van Buren, Suite 2610
Chicago, Ill. 60605

Clamshell Alliance News
62 Congress Street
Portsmouth, N.H. 03801

Critical Mass Journal
Critical Mass
P.O. Box 1538
Washington, D.C. 20013

Ecology Law Quarterly
School of Law, Boalt Hall
University of California
Berkeley, Cal. 94720

The Energy Daily
1239 National Press Building

Washington, D.C. 20045
Comprehensive coverage of all
   national energy news,
   energy policy, nuclear policy.

Environment
Helen Dwight Reid Educa-
   tional Fund,
in cooperation with
Scientists' Institute for Public
   Information
355 Lexington Avenue
New York, N.Y. 10022

Environmental Action
Environmental Action, Inc.
1346 Connecticut Ave. NW,
   Suite 731
Washington, D.C. 20036

Environmental Law
The Lewis and Clark Law
   School
Northwestern School of Law
10015 S.W. Terwilliger Blvd.

Portland, Oregon 97219
FAS Public Interest Report
Federation of American Scientists
307 Massachusetts Ave. SE
Washington, D.C. 20002
Groundswell
Nuclear Information and Resource Service (NIRS)
1536 16th Street NW
Washington, D.C. 20036
INFO
Atomic Industrial Forum
7101 Wisconsin Ave.
Washington, D.C. 20014
The Mobilizer
Mobilization for Survival
3601 Locust Walk
Philadelphia, Pa. 19104
Mother Jones
Foundation for National Progress
607 Market St.
San Francisco, Cal. 94105

Subscriptions:
1255 Portland Pl.
Boulder, Colorado 80302
Frequent articles on the nuclear energy front.

National Environmental Studies Project
Atomic Industrial Forum
7101 Wisconsin Ave.
Washington, D.C. 20014
NO (Nuclear Opponents)
Citizen's Energy Project
Box 285
Allendale, N.J. 07401
Not Man Apart

Friends of the Earth
124 Spear Street
San Francisco, Cal. 94105
Nuclear Engineering International
I.P.C. Electrical-Electronic Press, Ltd.
Dorset House
Stamford St.
London SE1 9LU England
Issue-length articles on new nuclear facilities and on particular aspects of nuclear technology.
Nuclear Industry
Atomic Industrial Forum
7101 Wisconsin Ave.
Washington, D.C. 20014
Nuclear News
American Nuclear Society
Oak and Catherine Sts.
LaGrange Park, Ill. 60525
Nuclear Safety
Nuclear Safety Information Center
Oak Ridge National Laboratory
P.O. Box Y
Oak Ridge, Tenn. 37830
Nucleonics Week
McGraw-Hill, Inc.
1221 Avenue of the Americas
New York, N.Y. 10020
International weekly summary of news of the nuclear industry.
Peace Newsletter
Syracuse Peace Council
924 Burnet Avenue
Syracuse, N.Y. 13203

People & Energy
Center for Science in the Public
    Interest
1757 S Street NW
Washington, D.C. 20009

The Power Line
Environmental Action Founda-
    tion
724 Dupont Circle Bldg.
Washington, D.C. 20036

The Progressive
Progressive, Inc.
408 West Gorham
Madison, Wis. 53703
Occasional articles on nuclear
    energy.

Science
American Association for the
    Advancement of Science

1515 Massachusetts Ave. NW
Washington, D.C. 20005

Scientific American
Scientific American, Inc.
415 Madison Avenue
New York, N.Y. 10017

Sierra Club Bulletin
Sierra Club
530 Bush Street
San Francisco, Cal. 94108

Supporters of Silkwood
317 Pennsylvania Avenue SE
Washington, D.C. 20003

Technology Review
Massachusetts Institute of
    Technology
Alumni Association
Cambridge, Mass. 02139
A pro-nuclear journal.

Organizations

American Friends Service Committee
Disarmament Conversion Program/AFSC
Terry Province
1501 Cherry St.
Philadelphia, Pa. 19102

Southeast Nuclear Transportation Project/AFSC
Chip Reynolds
P.O. Box 2234
High Point, N.C. 27261

Nuclear Facilities Network/AFSC
Pam Solo
1428 Lafayette St.
Denver, Colorado 80218

American Nuclear Society
Oak and Catherine Sts., LaGrange Park, Ill. 60525.
Professional organization for employees of the nuclear industry. Publication: *Nuclear News.*

Atomic Industrial Forum
7101 Wisconsin Ave., Washington, D.C. 20014.
Nuclear industry trade organization. Publications: *Nuclear Industry, INFO, National Environment Studies Project.*

Center for Science in the Public Interest
1757 S Street NW, Washington, D.C. 20009.
Citizen action on energy issues. Publication: *People & Energy.* Publications list available.

Citizen's Energy Project
1413 K Street NW (8th Floor), Washington, D.C. 20005.

Citizen's Energy Project
Box 285, Allendale, N.J. 07401.
Publication: *NO* (Nuclear Opponents).

Citizens for a Better Environment
59 East Van Buren, Suite 2610, Chicago, Ill. 60605.

193

Publication: *CBE Environmental Review*. Other publications on request.

Clamshell Alliance
62 Congress St., Portsmouth, N.H. 03801.
People opposed to the construction and use of nuclear power plants. Publication: *Clamshell Alliance News*.

Committee for Nuclear Responsibility, Inc.
P.O. Box 332, Yachats, Oregon 97498.
Education on the hazards of nuclear power and the alternatives to it. Publications available.

Common Cause
2030 M Street NW, Washington, D.C. 20036.
National citizen's lobby dedicated to making government accountable to the citizens.

Congress Watch
133 C Street SE, Washington, D.C. 20003.
Congressional lobby representing consumer interests, including the development of alternative energy sources.

Conservation Foundation
1717 Massachusetts Ave. NW, Washington, D.C. 20036.
A research, education, and information organization concerned with the quality of the environment.

Council on Economic Priorities
84 Fifth Avenue, New York, N.Y. 10011; 250 Columbus Avenue, San Francisco, Cal. 94133; 5443 South Kenwood, Chicago, Ill. 60615.
Information on the social responsibility of corporations, including the area of environmental impact. Publications available.

Council on Environmental Quality
722 Jackson Place, Washington, D.C. 20006.
Executive Office of the President. Advisory council on national environmental policy.

Critical Mass
P.O. Box 1538, Washington, D.C. 20013
Citizen movement organization for safe energy. Education on the problems of nuclear power. Publication: *Critical Mass Journal*.

Eastern Federation (of Nuclear Energy Opponents and Safe Energy Proponents)
317 Pennsylvania Ave. SE, Washington, D.C. 20003.
Clearinghouse for East Coast groups and alliances involved in anti–nuclear energy work.

Edison Electric Institute
90 Park Avenue, New York, N.Y. 10016.

Main association of investor-owned electric utilities. Weekly and bimonthly publications.

Educational Foundation for Nuclear Science, Inc.
1020–24 East 58th Street, Chicago, Ill. 60637.
Promotes study of the impact of science and technology on public affairs. Publication: *The Bulletin of the Atomic Scientists.*

Electric Power Research Institute
P.O. Box 10412, Palo Alto, Cal. 94303.
Research in the areas of nuclear power, energy analysis, and the environment. Publications available.

Energy Research and Development Administration (ERDA)
20 Massachusetts Ave. NW, Washington, D.C. 20545.
Federal agency for national energy research and development. Publications available through ERDA Office of Public Affairs.

Environmental Action, Inc.
1346 Connecticut Ave. NW, Suite 731, Washington, D.C. 20036.
National political lobby organization concerned with nuclear power. Publication: *Environmental Action.* Also publishes books and pamphlets.

Environmental Action Foundation
724 Dupont Circle Bldg., Washington, D.C. 20036.
Environmental research and resource organization. Publication: *The Power Line.* Also publishes books, manuals, pamphlets.

Environmental Action Reprint Service (EARS)
2239 Colfax, Denver, Colorado 80206.
Information on nuclear and other types of energy: books, articles, films, blueprints, posters. Write for catalog: *The Directory of Nuclear Activists.*

Environmental Defense Fund
475 Park Avenue South, New York, N.Y. 10016.
National environmental organization of lawyers and scientists to end "environmental degradation." Publication: newsletter.

Environmental Policy Center
317 Pennsylvania Ave. SE, Washington, D.C. 20003.
Research organization.

EPIC (Environmental Policy Information Center)
3 Joy Street, Boston, Mass. 02108.

Environmental Protection Agency (EPA)
Waterside Mall, 401 M Street SW, Washington, D.C. 20460

Federal agency for research, development, and enforcement of environmental policy. Publications available through EPA Office of Public Affairs.

Federation of American Scientists (FAS)
307 Massachusetts Ave. SE, Washington, D.C. 20002.
Public interest lobby. Research and education. Publications: *FAS Public Interest Report* and *Professional Bulletin*.

Friends of the Earth (FOE)
72 Jane Street, New York, N.Y. 10010; 124 Spear St., San Francisco, Cal. 94105; 620 C Street SE, Washington, D.C. 20003.
International conservation organization. Publication: *Not Man Apart*.

Green Mountain Post Films
P.O. Box 177, Montague, Mass. 01351.
Films available on nuclear power, nuclear resistance, the environment.

Institute for Energy Analysis—Oak Ridge Associated Universities
P.O. Box 117, Oak Ridge, Tenn. 37830.
Publications available.

Media Access Project
1910 N Street NW, Washington, D.C. 20036.
Public interest law firm. Information available on how to challenge utility advertising.

Mobilization for Survival
3601 Locust Walk
Philadelphia, Pa. 19104
407 S. Dearborn St. (Room 370)
Chicago, Ill. 60605

Southern California Alliance for Survival
5539 West Pico Blvd.
Los Angeles, Cal. 90019

Northern California Alliance for Survival
944 Market Street
San Francisco, Cal. 94102
Coalition of peace, environmental, and other groups challenging nuclear power. Publication: *The Mobilizer*.

Movement for a New Society
4722 Baltimore Ave., Philadelphia, Pa. 19143.
Works with local alliances in organizing and training for the anti-nuclear movement. List of publications available.

NARMIC (National Action Research on the Military-Industrial Complex)
1501 Cherry St., Philadelphia, Pa. 19102.
A project of the American Friends Service Committee. Resources for monitoring the nuclear front. Information available for local action groups. Fact sheets,

slide shows, publications available.

National Council of Churches of Christ

Energy Resource Consultant, Division of Church and Society, 475 Riverside Drive, Room 572, New York, N.Y. 10027.

Publications list available.

National Intervenors, Inc.

1757 S Street NW, Washington, D.C. 20009.

Coalition of environmental and citizen groups. Clearinghouse for public-interest opposition to nuclear power.

National Resources Defense Council (NRDC)

122 East 42nd Street, New York, N.Y. 10049; 917 15th Street NW, Washington, D.C. 20036.

Publications list available.

Nuclear Information and Resource Service (NIRS)

1536 16th Street NW, Washington, D.C. 20036.

Publication: *Groundswell.*

Nuclear Regulatory Commission (NRC)

Washington, D.C. 20555.

Federal agency for the regulation and licensing of commercial nuclear plants and facilities.

Nuclear Safety Information Center (NSIC)

Oak Ridge National Laboratory, P.O. Box Y, Oak Ridge, Tenn. 37830.

National clearinghouse for information on nuclear safety.

Publication: *Nuclear Safety.*

Public Citizen, Inc.

P.O. Box 19404, Washington, D.C. 20036.

Ralph Nader organization to support the work of citizen advocates. Publications available.

Public Interest Research Group (PIRG)

1346 Connecticut Ave. NW, P.O. Box 19312, Washington, D.C. 20036; NYPIRG, 5 Beekman Place, New York, N.Y. 10038.

Public-interest law group sponsored by Ralph Nader. Concerned with the issue of environmental quality. Information available on nuclear energy and its alternatives.

Resources for the Future

1755 Massachusetts Ave. NW, Washington, D.C. 20036.

Research and education on use and development of natural resources, energy, and improvement of environmental quality.

Scientists' Institute for Public Information

355 Lexington Avenue, New York, N.Y. 10022.

Publication: *Environment.*

Sierra Club
530 Bush Street, San Francisco, Cal. 94108.
National conservation organization. Publications include *Sierra Club Bulletin* and *Weekly National News Report*. Book list available.

Supporters of Silkwood
317 Pennsylvania Ave. SE, Washington, D.C. 20003.
Educational information related to the work of Karen Silkwood. Newsletter: *Supporters of Silkwood*.

Syracuse Peace Council
924 Burnet Ave., Syracuse, N.Y. 13203.
Publication: *Peace Newsletter*.

Task Force Against Nuclear Pollution
Box 1817, Washington, D.C. 20013.

Union of Concerned Scientists
1208 Massachusetts Ave., Cambridge, Mass. 02138.
Coalition of scientists and other professionals concerned with issues of nuclear safety and hazards, energy policy, proliferation. Publications list available.

War Resisters League
339 Lafayette St., New York, N.Y. 10012. Regional Offices: Southeast WRL, 108 Purefoy Road, Chapel Hill, N.C. 27514; WRL/West, 1360 Howard St. (2nd Floor), San Francisco, Cal. 94103.
Pacifist organization concerned with nuclear-power and nuclear-weapons resistance. Speakers and publications available.

World Information Service on Energy (WISE)
2e Weteringplantsoen 9, Amsterdam, Netherlands.
International research and reference source on the nuclear industry and the anti-nuclear movement. Newsletter and other publications.

Worldwatch Institute
1776 Massachusetts Ave. NW, Washington, D.C. 20036.
Research on global issues: energy, the environment, weapons proliferations. Publications list available.

NUCLEAR PLANTS THROUGHOUT THE WORLD

COUNTRY	TOTAL NUMBER	IN OPERATION	UNDER CONSTRUCTION
Argentina	2	1	1
Bangladesh	1	—	1
Belgium	8	4	2
Brazil	3	—	1
Bulgaria	4	2	2
Canada	25	8	9
Czechoslovakia	5	1	4
Federal Republic of Germany (West Germany)	32	12	15
Finland	4	1	3
France	35	11	22
German Democratic Republic (East Germany)	6	3	—
Hungary	4	—	2
India	8	3	5
Iran	4	—	2
Italy	9	3	4
Japan	28	11	13
Korea	3	—	3
Luxembourg	1	—	—
Mexico	2	—	2
Netherlands	2	2	—
Pakistan	1	1	—
Philippines	2	—	—
Poland	1	—	—
Rumania	1	—	—
South Africa	2	—	—
Spain	16	3	9
Sweden	12	6	5
Switzerland	7	3	2
Taiwan	6	—	6
UK	20	14	6
USA	222	70	79
USSR	41	25	16
Yugoslavia	1	—	—
Total	518	187	214

Index